WAYS OF COMMUNICATING

The Darwin College Lectures

Ways of communicating

EDITED BY D H MELLOR

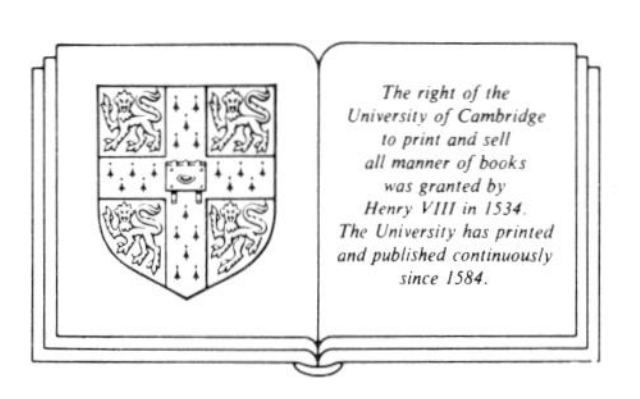

CAMBRIDGE UNIVERSITY PRESS
CAMBRIDGE
NEW YORK PORT CHESTER
MELBOURNE SYDNEY

Published by the Press Syndicate of the University of Cambridge
The Pitt Building, Trumpington Street, Cambridge CB2 1RP
40 West 20th Street, New York, NY 10011, USA
10 Stamford Road, Oakleigh, Melbourne 3166, Australia

First published 1990

Printed in Great Britain at The Bath Press, Avon

British Library cataloguing in publication data

Ways of communicating. – (The Darwin College lectures).
1. Communication
I. Mellor, D. H. (David Hugh), *1938–* II. Series
302.2

Library of Congress cataloguing in publication data

Ways of communicating: the Darwin College lectures/
edited by D.H. Mellor.
p. cm.
1. Communication.
I. Mellor, D. H. II. Title: Darwin College lectures.
P90.W38 1990
302.2–dc20 90–1469 CIP

ISBN 0 521 37074 4 hardback

WV

CONTENTS

CONTRIBUTORS

JOHN ALVEY
Former Engineer-in-Chief
British Telecom

PROFESSOR HORACE BARLOW
Kenneth Craik Laboratory of Physiology
University of Cambridge

PROFESSOR PATRICK BATESON
Sub-Department of Animal Behaviour
University of Cambridge

PROFESSOR NOAM CHOMSKY
Department of Linguistics and Philosophy
Massachusetts Institute of Technology

PROFESSOR ALEXANDER GOEHR
Faculty of Music
University of Cambridge

DR P N JOHNSON-LAIRD
Department of Psychology
Princeton University

DAVID LODGE
Honorary Professor of Modern English Literature
University of Birmingham

PROFESSOR D H MELLOR
Faculty of Philosophy
University of Cambridge

JONATHAN MILLER
Artistic Director of the Old Vic Theatre, London

Introduction
What is communication?

P N JOHNSON-LAIRD

In *The Expression of Emotions in Man and the Animals*, Charles Darwin wrote 'The power of communication between the members of the same tribe by means of language has been of paramount importance in the development of man; and the force of language is much aided by the expressive movements of the face and body' (p. 354). Darwin was right; but like most nineteenth-century thinkers he took for granted the notion of communication. It seemed to be a self-evident capacity that confers an evolutionary advantantage on those species that possess it:

> You see a sabre-toothed tiger.
> You warn me.
> I am thereby able to avoid it and to survive.
> End of story.

Yet, communication is not simple. In fact, it is profoundly complicated; and the revelation of its hidden complexity is one of the great discoveries of the twentieth century.

One sure sign of this complexity is our ignorance. We still do not know, for instance, how the human auditory system recognises speech, and so we cannot built a machine that will convert speech into typewriting. (Existing devices fall far short of human ability.) We do not have a complete grammar for any natural language; we do not have a comprehensive account of how sentences convey meanings. We communicate emotions, as Darwin said, but we are still not certain what emotions are. And, when it comes to the full interplay of all the factors at work in everyday conversation, our theoreticians are hopelessly lost. Scholars in Darwin's time did not know that they did not know. As modern researchers have begun to explain communication, they have rectified this second-order ignorance. But the complexity of the subject is reflected again in the need for interdisciplinary research.

The present book surveys communication from the standpoints of both the arts and the sciences. Its contributors are, in order of their appearance: a physiologist and an ethnologist, a linguist and a philosopher, a novelist and a man of the theatre (who is also a neuropsychologist), a composer and an engineer. Communication is their business or their object of study, or both, and each of them presents a coherent view of one aspect of it. None of them, however, was asked to analyse the concept of communication itself. My aim in this introduction is accordingly to provide such an analysis in order to help the reader to see how the various parts fit together into a unified picture. In short, I shall try to answer the question left unanswered by Darwin: what is communication?

Knowledge seldom advances by virtue of *a priori* definitions, and so I will proceed, not by way of definition, but by an analysis of some case histories. The first cases set the scene, and allow a line of demarcation to be drawn between communication and other sorts of causal influence. Later cases will distinguish between different varieties of communication.

The moon exerts a causal effect on the tides, but it does not, except metaphorically, communicate with them. Communication is a matter of causal influence too, but it calls for something more. In particular, a communicator has a message to transmit, whereas the moon has none. The notion of a message seems equally problematical, but the next case will help to sharpen up our intuitions about it.

When a fire ant returns to the nest from an abundant food source, other workers are able to go to the food by tracking a chemical substance – a so-called 'pheromone' – laid down in a trail by the returning worker. When the food runs out, workers return to the nest without laying down the trail. Other pheromones act as insect sexual attractants, alarm signals, and probably play a part in mammalian behaviour.

Sociobiologists, such as Edward O. Wilson, routinely refer to pheromones as 'chemical communications', and it is tempting to suppose that the fire ant has a message to communicate, namely: 'Food!' It may lay down its trail, however, solely as a physiological consequence of tasting food, and the behaviour of the other workers may likewise be a direct physiological consequence of their detection of the pheromone. The ants, in effect, respond to the 'odour' by following it; and its detection may elicit nothing apart from this response. Certainly, there is no need for the ant's nervous system to construct an internal representation of some rudimentary 'idea' of food that lights up in the insect's mind as soon as it smells the pheromone. Indeed, the use of a trail counts against such an internal representation: you do not need a map if you are travelling by rail. A biologist observing the ants can talk of 'chemical communication', but the ants themselves may form no internal representation of food, and their tracking behaviour may be governed by factors outside their representation of the world. Whether one still chooses to refer to the process as 'communication' is a matter of lexicographical convenience. The important point is that it lacks a component: there is no message, only a causal influence.

When a foraging bee returns to the hive, she disgorges the food she has gathered, and then performs a dance on the vertical honeycomb. The tempo of the dance, as Karl von Frisch showed, is related to the distance of the food source: the closer the food, the faster the dance. Likewise, as Patrick Bateson explains in chapter 2, the angle from the vertical to the dance's main trajectory depends on the angle from the direction of the sun to the direction of the food. The workers who attend the dancer (and who also pick up pheromonal cues) are thus able to fly almost unerringly to the source of the food.

The bee's dance conveys about the same amount of information as the ant's pheromonal trail. Yet, there is a subtle difference between the two cases. Granted that a message depends on an internal representation of a state of affairs, the dancing bee does have a message to communicate. The evidence comes from one of von Frisch's ingenious experiments. He placed a hive on one side of a mountain ridge and food on the other side, and so the bees were forced to fly a dog-leg route. When they returned to the hive, they did not perform a dance indicating the direction in which they had flown – neither their initial course nor their change of course in midflight. Instead, they danced a 'beeline' from hive to food directly through the mountain rather than around it. That they were able to work out this direction establishes that their nervous system constructs a representation of the world. This representation may encode only two dimensions because there is no sign for 'up' or 'down' in the bees' dance, and they cannot communication the vertical location of food. Nevertheless, bees have a message; it guides their dance, and so they succeed in communicating it.

The bees' dance is symbolic. Its primary purpose is to symbolise the distance and direction of food. Like all symbolic behaviours, it is arbitrary in the following sense: quite different symbolic conventions could have been embodied in the dance without affecting any other facet of the bee's life. Thus, the distance of the food could have been represented by the distance traversed in the dance. The notion of an arbitrary relation between a symbol (or an element in symbolic behaviour) and what it symbolises is obvious in the case of natural language: different linguistic communities use different words for the same things. Indeed, it was the Swiss linguist and founder of structuralism, Ferdinand de Saussure, who first emphasised the arbitrary nature of the symbolic relation.

A sceptic might argue that the fire ant's trail depends equally on an arbitrary choice of pheromone: natural selection could have contrived to use a different one. But, if the detection of a pheromone controls the ant's behaviour by direct physiological means, the choice is not arbitrary. Change the pheromone, and much else besides in the ant's physiology must also be changed. The point is still clearer in another case.

When you ingest glucose, it is either used immediately as a source of energy (in the form of a substance known as ATP, adenosine triphosphate) or it is stored for future use (in the form of glycogen). Its fate depends on whether or not you need energy at the moment. One can imagine a robot that used internal symbols to represent its available energy; and the choice of the particular symbols would be arbitrary in just the Saussurian sense that concerns us. But, although a feedback system in the body controls the fate of the ingested glucose, it is not in the least symbolic. The enzymes that govern the storage or breakdown of glucose are themselves simply turned on or off by the actual amount of ATP in your body. Nothing *represents* the amount of available energy: the amount itself controls the system. If ATP were changed for some other substance, then the entire physiology of muscles would have to be changed, too. Pheromones may be similarly embedded in a physiological feedback system where their bio-chemical properties have consequences beyond their quasi-symbolic role.

The picture of communication that has so far emerged calls for the communicator to construct an internal representation of the external world, and then to carry out some symbolic behaviour that conveys the content of that representation. The recipient must first perceive the symbolic behaviour, i.e. construct its internal representation, and then from it recover a further internal representation of the state that it signifies. This final step depends on access to the arbitrary conventions governing the interpretation of the symbolic behaviour.

One complication is that communication can be a one-sided affair. When you smell the odour of newly baked bread, you may recognise it, and, like a fire ant, track down the bakery by following the smell. The bakery did not communicate with you, because its smell is a by-product, not a message. Nonetheless, you formed an internal representation of the bread just as though you had received a message from it. Conversely, when you soothe a fractious infant with calming words and gentle caresses, you communicate a message – one that any adult third party would easily understand – and yet its effect upon the baby may be a result of its physiological accompaniments rather than its symbolic content.

The study of animal communication is riddled with spurious cases in which what seems to be full-blooded communication turns out to be a one-sided affair. An animal appears to respond to a symbolic message but in fact has merely become conditioned to an unconscious and involuntary cue from its trainer that is correlated with the content of the message. Similarly, apparent communication *between* animals may depend on nothing more than a mutually one-sided detection of the other's response to the situation. For example, one dolphin seems to communicate to another which of two different shapes has been presented to it – with the result that both of them are rewarded with food when the second dolphin makes the appropriate response. One must be careful, however, not to read too much into this trick. It can also be performed, as Robert Boakes has shown, by pigeons. At the start of the experiment, the 'receiver' has a bias towards pecking one of two keys, which ensures that both birds receive food whenever this response happens to be correct. The 'communicator' therefore learns that one particular stimulus leads to food, and so behaves differently when it occurs. The 'receiver' then learns to make the preferred response only when the 'communicator' behaves in this anticipatory way. Ultimately, the birds may mimic the entire process of communication, and yet neither has a message to communicate. Their behaviour is not symbolic: two one-sided forms of communication have dovetailed to simulate the process.

The analysis of communication as the symbolic transmission of a mental representation applies pre-eminently to the communication of a simple propositional content – from the location of honey to a warning about a sabre-toothed tiger. It must be modified, however, for most richer forms of communication, both those that convey information about the communicator's emotional state, and those that depend on the full resources of natural language.

The life of a social mammal such as a rat (or a human being) is governed by emotions. They probably serve, as Keith Oatley and I have argued, a twofold communicative function. First, the perception of certain events elicits an *internal* emotional signal. This signal prepares the individual for a general course of action appropriate to the situation. It elicits a more flexible response than the fixed responses of insects; it operates more rapidly than the decoding and evaluation of complex symbolic messages. The events that trigger these signals pertain to such matters as the making of attachments, hostility towards rivals, and co-operation against predators. Second, as Darwin emphasised, the emotion spills over into the expressive behaviour of the individual, and in this way communicates itself to others by an *external* signal. The important situations in interacting with others can indeed be mapped into a small set of basic emotions that appear to be common to all social mammals (happiness–sadness, anger–fear, desire–disgust).

Emotions arise from the perception of events but their communication normally conveys, not the content of the resulting internal representation, but the individual's reaction to that content. The perception of a predator, for instance, creates fear within the individual and prepares it for fight or flight. Its alarm call communicates the emotion to other members of the group. Such cries are ritualised symbolic behaviours that may no longer serve any purpose other than to communicate the emotion. The primary content of the message is a contagious emotional state, and so the recipient is in turn infected by an emotion – the same one in a co-operative situation, and an antagonistic one in a confrontation. The external signal creates the internal signal.

Human beings can communicate in more complex ways than any other animal, and pre-eminently by using natural language. Whatever its origins, and whatever the other purposes for which it is used – for self-expression, for the externalisation of thought – it is richer than any other known symbolic system. To paraphrase George Miller, the founder of modern psycholinguistics, all human groups speak a language, and all human languages have a grammar and a lexicon. The lexicon always has words for dealing with space, time, and number, words to represent true and false, and words for communicating logical relations. The grammar contains principles governing phonology and principles governing syntax. The phonology always contains vowels and consonants, which can be described in terms of a set of features characterising the limited set of humanly possible sounds. The syntax always contains principles governing the intonation of utterances, and the hierarchical structure of phrases and sentences.

The power of language derives from three principal factors. First, the lexicon provides speakers with a large repertoire of individual symbols (words). Second, the grammar enables these symbols to be combined into an unlimited number of distinct symbolic messages (sentences). Third, these messages are not under the immediate control of the local environment. They can be intentionally used to refer to other states of affairs including those that are remote, hypothetical, or imaginary. Human beings can tell stories (in both senses of the word). They can give one another instructions and requests. They can do things with words.

Some chimpanzees have been taught the American sign language for the deaf, and others have been taught to manipulate symbolic shapes, and these animals have been able to 'talk' in a rudimentary way about their needs and about their immediate environment. Their accomplishments, however, are controversial. Whether they can master the syntactic and semantic power of a natural language are questions that have yet to be answered to everyone's satisfaction.

In human communication, there is an intricate interplay between language and other communicative modes. When we talk to one another, we are fully engaged in the communication of content, attitude, emotion, personality. Tone of voice, for example, is an important modulator of the literal content of what we say. Once upon a time Stalin read out in public a telegram from Trotsky: 'You were right and I was wrong. You are the true heir of Lenin. I should apologize. Trotsky.' According to Leo Rosten, a Jewish tailor then stepped from the crowd and explained to Stalin how he ought to have read the message:

> You were right and I was *wrong*? *You* are the true heir of Lenin? *I* should apologize???!!

Although we all immediately recognise the ironic import of this intonation, few of us consciously know how our vocal apparatus achieves this or any other of its effects. Phoneticians and psycholinguists, of course, can tell us something of what we are doing. But, our spontaneous use of language depends on profoundly unconscious processes – not those that a psychoanalyst might hope to uncover from a patient's free associations on a couch, but those that make our conscious experience possible. They can never enter consciousness, because they create its contents: you cannot pick up your own introspective process by its bootstraps.

The unconscious processes that enable us to speak and to understand a language operate at all levels – from the identification of individual speech sounds to the rapid automatic inferences that we make in order to flesh out the explicit content of utterances. These inferences are vital because the use of language is almost always dependent upon context and particularly the background assumptions that speaker and hearer share. Discourse often makes sense only if you are privy to these assumptions. On the night that Mrs Thatcher won her first election (in 1979), the former leader of the Tories, Mr Edward Heath, was interviewed on television. The key part of the interview went as follows:

> *Interviewer (Sir Robin Day)*: I think you know the question I am going to ask. What is your answer?
> *Heath*: We'll have to wait and see.
> *Interviewer*: Would you like to?
> *Heath*: It all depends.

If you do not know the question that the participants (along with the country as a whole) had in mind, the exchange is a self-parody of English diffidence. The interviewer's question, 'Would you like to?' is particularly problematical because its ellipsis depends on taking the words of the question for granted. It was, of course: 'Will you serve in Mrs Thatcher's cabinet?'

Each new utterance in a conversation changes the context, and thus can play a role in the interpretation of the subsequent discourse. And context is an indispensable part of nearly all human communications. It is almost impossible outside mathematics to frame a sentence that does not depend on context and background assumptions for its interpretation. Attempts to devise 'eternal' sentences, such as:

> At 2 p.m. on 12 October 1492, Christopher Columbus sighted America.

fail at once for a Martian who wants to write a postcard home, because they depend on assumptions about calendars and the measurement of time. Indeed, so dependent is interpretation on context that some philosophers and linguists have argued that no scientific theory of its role will ever be possible. Noam Chomsky remarks (in chapter 3): 'virtually any information and strategy might be relevant to determining what a presented utterance means, or what its speaker may have had in mind'.

A more pessimistic doctrine can be traced back from certain literary critics to Jacques Derrida, and ultimately to that *éminence noire* of continental philosophy, Martin Heidegger: the quest for the correct interpretation of a communication is futile. Human discourse is not the transmission of information: there is no world independent of us: the only world is the one that we create through our language: and so on and on, until eventually this deconstructive doctrine deconstructs itself. How, you may wonder, were you able to warn me about the sabre-toothed tiger?

In fact, people can and do communicate successfully. A speaker perceives a state of affairs, that is, constructs a mental model representing it. The speaker intends to communicate certain aspects of this situation to a listener, and so, taking into account common knowledge, utters some appropriate words. The listener perceives these words and, again taking into account common knowledge, is able to grasp the content of the speaker's immediate communicative intention. The listener constructs a mental model representing the relevant features of the original situation. A message may then pass in the opposite direction, and, as a result, the two participants may share a mutual knowledge that the act of communication has been successfully consummated. Perhaps paradoxically, the best evidence for the existence of a correct interpretation of discourse is the fact that failures of communication occur, and are known to occur. If no discourse had a true interpretation, such failures could neither occur nor be rectified. Of course, there is no end to the process of recovering speakers' intentions – why they chose to communicate this or that information. And a text does not talk back, and hence as its author's background assumptions fade into obscurity so its interpreters are free to project ever wider and ever more idiosyncratic readings onto it. But that is as much a fact about human psychology as about communication. Deconstructionism confuses the admitted difficulty of recovering the communicator's intentions with a wholly independent question: do communications ever have a correct interpretation?

Discourse about imaginary states of affairs is, as I have mentioned, a human prerogative. Although the question of what fictions mean, and what fictional referring expressions, such as 'Hamlet's mother', refer to may be philosophically problematical, understanding such discourse differs in no discernible way from understanding factual discourse. Both call for the construction of mental models of the states of affairs and events that the discourse describes. The only difference is that in one case the model purports to correspond to the world, and in the other case it does not. In one case, you may sensibly ask: 'Is that true or false?', and, as Hugh Mellor points out in chapter 4, the answer may materially influence your ability to cope with the world.

At some point in the evolution of human beings communication became an end in itself. Consider the following brief poem:

> Swiftly the years, beyond recall.
> Solemn the stillness of this spring morning.

The late Sir William Empson commented on this translation of an ancient Chinese poem:

> Lacking rhyme, metre, and any overt device such as comparison, these lines are what we should normally call poetry only by virtue of their compactness; two statements are made as if they were connected, and the reader is forced to consider their relation for himself. The reason why these facts should have been selected for a poem is left for him to invent; he will invent a variety of reasons and order them in his own mind. This, I think, is the essential fact about the use of language.

A more extreme view is Paul Valéry's aphorism: 'Poetry is made from words, not ideas.' That, perhaps, is an exaggeration, but it is an antidote to the notion that genuine poetry could be composed without a concern for the words themselves. Poetry may have no significant message to communicate other than itself; and music too – particularly in the European tradition – may have no meaning beyond itself. Yet there appears to be a component common to all these 'messages': they elicit an emotional response from those individuals attuned to them. The final complication in the analysis of communication is therefore that for human beings the symbols themselves, rather than their interpretation, may come to be the important component of the message.

In this introduction, I have laid out an informal analysis of the main varieties of communication from the simplest of propositional messages to those whose significance is problematical. My aim has been to prepare the ground for the subsequent chapters, which I will now introduce.

A prerequisite for genuine communication, I have argued, is the ability to construct internal representations. This process depends itself on both physical causes and internal communications. Vision, for instance, begins with the optical process of focussing light on a light-sensitive surface, and the biochemical conversion of light into nerve impulses. These nerve impulses are the beginnings of a communicative chain within the brain. In chapter 1, Horace Barlow examines the nature of this chain, and the fate of nerve impulses emanating from cells in the retina.

An organism that constructs internal representations may be able to communicate them to others, and such a system of communication is likely to serve a purpose in the life of the organism. The nature of that purpose – the potential reason that the system evolved – is analysed in Patrick Bateson's account of animal communication in chapter 2. Some biologists believe that the aim of animal signals is the Machiavellian manipulation of others, but Bateson establishes that many signals do accurately communicate the internal state of the signaller. It may be advantageous to be deceptive in the rituals of combat; it is hardly so in the rituals of courtship. A plausible extrapolation from animal communication suggests that natural language evolved to aid human communication. The extrapolation, however, is open to doubt. Bateson shows that certain cognitive precursors needed for language exist in other animals, but no one has yet established the existence of genuine syntactic ability outside human beings.

Noam Chomsky, who has revolutionised the modern study of linguistics, argues in chapter 3 that language is '"beautiful" but not "usable"'. There is no evidence, he says, that it evolved for ease of use or ease of acquisition. Its function may be to externalise thought rather than to communicate; speech may be a by-product of the evolution of a system of breathing that allows human beings to run with their mouths open; syntax, as Chomsky claims, may be a by-product of the evolution of a complex brain. Yet people can communicate provided that their internalised languages – their lexicons and grammars – are close enough. How they come to possess this internalised knowledge is a question that Chomsky addresses in his chapter.

One purpose of communication is to enable us to experience the world by proxy, and this function of communication is taken up by Hugh Mellor in chapter 4. Truth and falsity, he argues, are properties of both beliefs (consciously accessible internal representations) and those symbolic behaviours, including utterances, that communicate beliefs. Intentional communication depends, of course, on an awareness of such a belief. As a result of observation, you acquire a true belief, and then act so as to communicate it to me. As a result of my observation of your behaviour, I come to adopt the same belief because I believe that you believe it. Knowledge by description is thus, for Mellor, an indirect form of knowledge by acquaintance: communication is indirect observation.

Human communication normally starts and ends with conscious messages, but the intervening mental processes are almost wholly unconscious. They are, as I remarked earlier, largely the same for both fact and fiction. Every night we are all engaged in the construction of narratives in the form of vivid hallucinatory images. We have normally forgotten our dreams the morning after. No one knows the purpose of these 'communications to ourselves', but one plausible conjecture is that they keep us entertained whilst the brain has temporarily paralysed our bodies to keep us out of harm's way at dead of night. The novel raises the construction of narratives to linguistic art. It is, as David Lodge asserts in chapter 5, a form of communication, but a problematic one. The comprehension of action and dialogue is little different from their comprehension in the description of real events – otherwise, how should we be as amused by the antics of Maurice Zapp as, say, by those of Norman Mailer? But, because the text cannot answer back, the author's original intentions cannot be recovered for certain (not even by the author). What lies behind the creation of a novel, and what its deepest significance, if any, might be, are deep puzzles. Like the interpretation of dreams, they are topics for endless speculation.

The different components of human communication are often not so much separate threads conveying separate messages but are rather woven together into a single vivid fabric. This theme underlies Jonathan Miller's account in chapter 6 of non-verbal communication. He, too, distinguishes between the unconscious betrayal of information – one-sided communication, as I referred to it earlier – and the deliberate communication of information by gesture and action. It is easy to imagine that such behaviours convey a rudimentary emotional accompaniment to the principal business in hand: the use of language to convey beliefs. Yet, Miller argues convincingly that this view is untenable. Following Erving Goffman, he shows that non-verbal communication must be interpreted in relation to the norms of our social conventions. In public places, he said, our behaviour transmits a stream of signals intended to communicate to others our right to be where we are, doing whatever we are doing. The message may have no unique propositional content, but it is no mere residue of the rudimentary symbolic systems of animal communication. It calls for the meta-cognitive ability to think about what others may be thinking – to construct mental models of their mental models – and it reflects some of our deepest moral concerns about what we are, and what we may seem to be, as autonomous moral agents.

For many individuals, music is the profoundest form of communication. But what does it communicate? The analogy between music and language, as Alexander Goehr remarks in chapter 7, breaks down over this question of semantics. Nevertheless, much of the communications industry is given over to the production and dissemination of music because nearly everyone enjoys listening to it. Why? Undoubtedly, because music moves the emotions. But this answer replaces one puzzle with two: how does music communicate emotions, and why do we enjoy having our emotions stirred in this way? No one knows. Goehr leaves open the question of whether its emotional effects are a result of innate factors or cultural conventions. What he shows, however, is that the significance of music depends on its context, and that perhaps its most authentic understanding comes only when one participates in a communal performance.

Nowadays, the word 'communication' is apt to elicit images of satellites circling the earth, beaming their signals to and fro. Such satellites were first mooted by the novelist, Arthur C. Clarke. In a delightful vignette in Clarke and Kubrick's film *2001*, a scientist phones home from an orbiting space station. This event has now been made possible by the panoply of high technology that John Alvey describes in the final chapter of the book. The scientist asks his daughter what she wants for her birthday. And she says: 'a telephone'.

FURTHER READING

Darwin, Charles. *The Expression of Emotions in Man and the Animals*. London: Murray, 1872.
Empson, W. *Seven Types of Ambiguity*. London: Chatto and Windus, 1930 (3rd edn, Harmondsworth, 1961).
Frisch, Karl von. *The Dancing Bees*. London: Methuen, 1954.
Johnson-Laird, P. N. *The Computer and the Mind*. London: Fontana; Cambridge Mass: Harvard University Press, 1988.
Miller, George A. *The Psychology of Communication: Seven Essays*. Harmondsworth, 1968.
Saussure, Ferdinand de. *Course in General Linguistics*. London: Peter Owen, 1960 [Originally published, 1916].
Wilson, Edward D. *The Insect Societies*. Cambridge, Mass: Belknap, Harvard University Press, 1971.

1

Communication and representation within the brain

HORACE BARLOW

This chapter is about the way one part of the brain communicates with a different part of the brain. How, for example, does my left cerebral hemisphere, which occupies the left half of my skull, talk to the right cerebral hemisphere, which is about three inches away? I think you will agree that, if we knew about this, we would know the answers to a lot of interesting questions. Do the messages that pass from one side of the brain to the other use symbols like the words of our ordinary language? Is there a grammar? Does my left brain tell the truth to my right brain, or can it lie to it? I shall not be able to answer these questions, but hope to convey some of the ideas that neuroscientists are currently pursuing in search of answers.

Two of the things we do know about communication and representation within the brain are, first, that they are brought about by *nerve cells*, and, second, that the *neocortex* is the most interesting bit of the brain to consider since it is the part whose great expansion distinguishes the primates from other mammals, and makes humankind pre-eminent among primates. It is a paired structure which occupies most of the *cerebral hemispheres* and is also called the *neopallium*. Sixty years ago C. J. Herrick, the American comparative anatomist, dubbed it the 'organ of civilisation', and I have set myself the task of seeing how far our scientific knowledge of nerve cells might earn the neocortex this grandiose title: Would these nerve cells, as actors, be able to perform the play, 'Civilisation'?

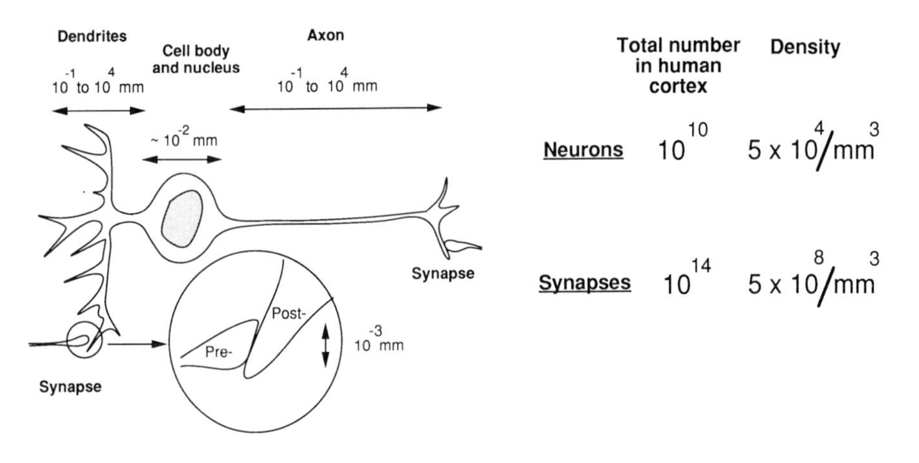

Figure 1 Diagram to show the size of a neuron and its components, and their ranges of variation. Inset shows a single synapse. The numbers at the right give an impression of the complexity of the computational system in our heads.

NERVE CELLS

Varieties Nerve cells, or neurons, come in all sorts of shapes and sizes and perform a great variety of functions. For example, pinching the skin excites nerve terminals there, and nerve fibres propagate this excitation from the skin to the spinal cord and brain, where the messages give rise to the sensation of pain. Other neurons conduct impulses from the spinal cord to muscles and cause them to contract; so the pain relayed by the first neurons might cause a limb to be moved by the second set of neurons. There are also many nerve cells within the brain that connect one part to another, and these are the ones I shall mainly talk about.

Size and number Figure 1 shows the parts and dimensions of a much simplified neuron. The distance of the *cell body* from the *dendrites* is extremely variable. Within the brain it is usually only 1 mm or less, but the cell bodies of the neurons that bring information to the brain from the skin lie close to the spinal cord, so in a whale or dinosaur the dendrites could be 10,000 mm long, and this also applies to the *axons* of the motor fibres, which carry commands from the spinal cord telling the muscles to contract. Cell bodies are more constant in size – from about 1/100 to 1/10 mm.

Rather complicated interactions can occur at the *synapses* that connect

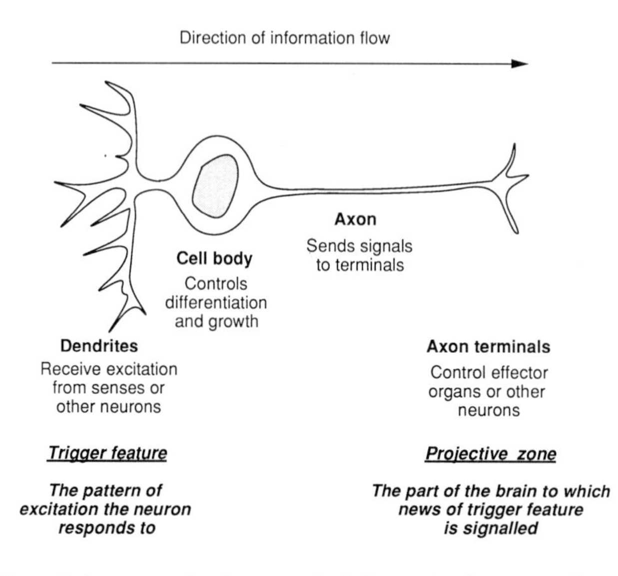

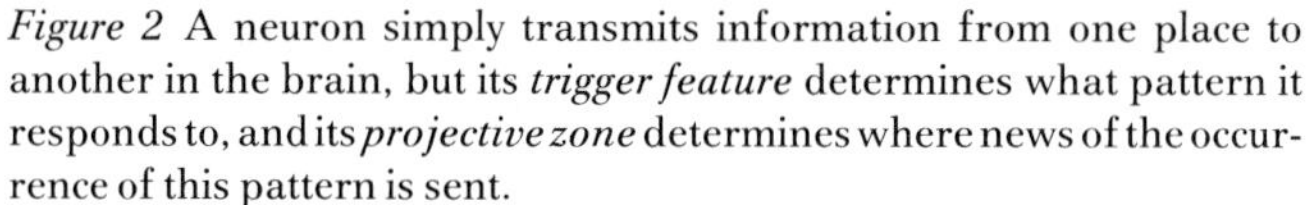

Figure 2 A neuron simply transmits information from one place to another in the brain, but its *trigger feature* determines what pattern it responds to, and its *projective zone* determines where news of the occurrence of this pattern is sent.

one cell to another, so the dimensions of a synapse are also of interest. They are of the order of 1/1000 mm, and each neuron may have 10^3 to 10^5 such synapses from other neurons on its dendrites and cell body.

In the table at the right of Figure 1 I have given rough figures for the numbers of neurons and synapses in the human neocortex. As you see, these are very large, and one possibly gains better intuitions from the densities at the right. There are more than 200,000 cubic millimetres of neocortex, and every cubic millimetre contains at least 50,000 neurons with 500 million synapses; these figures may give you some idea of the complexity of the computer we have in our heads.

Information flow Figure 2 shows what such a simplified neuron does. Its dendrites pick up information from other neurons, and the axon and its terminals pass this on to other cells. The cell body with its nucleus controls the growth and maintenance of the cell. The axon transmits information by means of electrical impulses, and it is because these cause current to flow in the media around the cell that one can monitor the impulses, and hence the information flow, by means of a micro-electrode placed close to it.

Simply transmitting information from input to output may seem a trivial task, but the neuron has two opportunities to be discriminating in performing it. First, the details of the synaptic connections from other neurons on to the dendrites determine what a neuron responds to; think of them as forming a *lock* which is only opened by a *key* in the form of a particular pattern of activity in the neurons which make contact with it. This is sometimes called the *trigger feature*, and I shall shortly show you examples.

The second crucial feature of neurons is the ability to relay this information to a particular destination in the rest of the brain, its *projective zone*. So each neuron responds to a particular pattern of activity in the cells that connect to it, and when this pattern occurs it signals the news to a group of cells lying in another part of the brain. A third key element of a neuron's computing power is its *modifiability*, and I shall consider this after giving examples of trigger feature and projective zone.

Retina I am going to illustrate these ideas by a brief discussion of neurons in the retina of the rabbit. I have selected them partly because I was involved in their investigation,[1] but also because it is a relatively simple system that does happen to illustrate these two concepts rather nicely.

The vertebrate retina is an outlying part of the brain. The light in the image formed by the lens actually passes through the neural layers of the retina (which are almost transparent) to reach the photoreceptors where it is absorbed. The ganglion cells are the neurons which communicate with the brain by propagating impulses up their axons, while their dendrites detect patterns of excitation in the photoreceptors, to which they connect through intermediate cells. For present purposes we need not worry about the complexity and the names of the various components in the retina if only we can answer the following question: What does a single one of these ganglion cells tell the brain about the image falling on the photoreceptors to which it connects? It turns out that this does not have a simple answer, for there are many different kinds of ganglion cell (as anatomists have long known), and when one records from them one finds that they carry different types of message. Let us look at some of them.

Receptive fields Figure 3 shows responses from two of the commonest types of cell which are found in most vertebrate retinae. The first and third columns illustrate light stimuli, the second and fourth show about one second's worth of the electrical responses from each cell; the electrical impulses appear as vertical deflections on these records. For the top row,

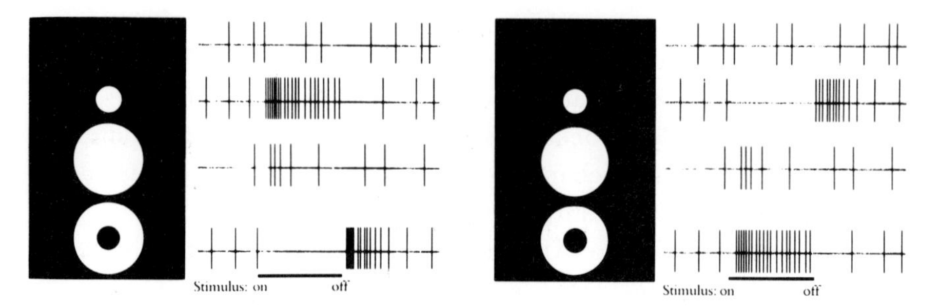

Figure 3 Responses of an *On-centre* (left) and *Off-centre* (right) ganglion cell from typical mammalian retina. The left part of each half shows the stimulus configuration which produced the pattern of impulses shown in the right part. The periods when the light was on are shown by the bars below (from Hubel 1988: see Further Reading).

no stimulus was applied, but nevertheless the cells give a slow and irregular sequence of these impulses. Now all retinal ganglion cells have, somewhere within the visual field, a region called the *receptive field* where light causes a change in the number of impulses, and as you would expect this region is situated within the field of view of the eye at the position where a light spot illuminates the retina underlying the ganglion cell.

First look at the second row of the left half. If you put a spot of light right in the centre of this cell's receptive field you get the response shown in the second column: when the spot goes on (see stimulus trace at the bottom) you get a burst of impulses, so it is called an 'on-centre unit'. At the right are shown responses from the other type: here the burst of impulses occurs when the spot of light goes off, so it is called an 'off-centre unit'. Surprisingly, if you use a large spot that covers the whole of the receptive field, you get very little response in both cases, as shown in the third row. The reason for this becomes clear if you use an annular stimulus; you then get a response at the opposite phase to that for a central stimulus, and there is antagonism between these two regions, so when you illuminate both together there is little or no response; they cancel each other out.

It is not difficult to see what such neurons are telling the brain. The 'on' type is saying that the centre of its particular region of the visual field is brighter than the rest, while the 'off' type is saying that it is darker. The 'keys' or trigger features required to 'unlock' these two types of cell are rather simple – it is just a matter of applying a small stimulus in the right place, and giving it the correct sign, an increase for one and a decrease for the other.

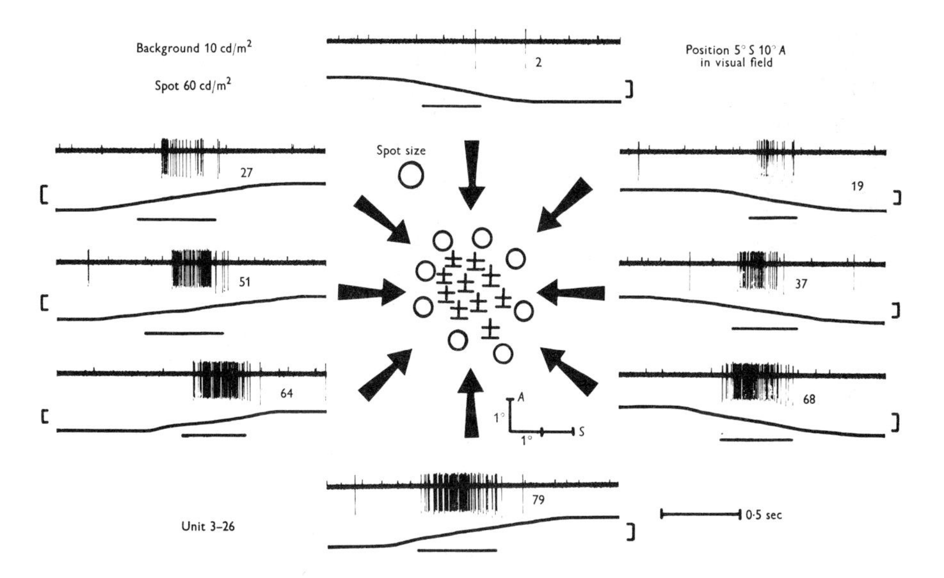

Figure 4 This ganglion cell from a rabbit retina gave responses when a stationary spot was turned on (+) or off (−) anywhere within the area surrounded by O's. The receptive field does not enable one to predict the responses to *movements* of a spot in the directions of the arrows; these are shown round the outside of the figure. The *trigger feature* for this cell is movement upward through its receptive field (from reference 1).

Trigger features The next trigger feature is rather more complicated, and is illustrated in Figure 4. The centre shows a map of the receptive field of a particular ganglion cell obtained using a stationary spot. In the convention used here, a plus sign signifies that a response was obtained at onset at that position, a minus sign that a response occurred at offset, and 0 means that no response was obtained there or outside that position. So this cell signals both onset and offset of a small spot over a small region of the visual field. But when you explore the same region of the visual field with a *moving* spot you get a result that you could not have predicted from the responses with stationary spots. For an upward movement the massive response shown below is given, while for a downward movement only two impulses result, as shown at the top. For other directions the responses are intermediate.

The message conveyed to the brain from one such cell is a bit ambiguous: it is saying either that a stationary spot is going on or off in a particular region, or that something is moving upwards in that region. However, in a natural environment a sustained response is fairly unambiguous and must indicate

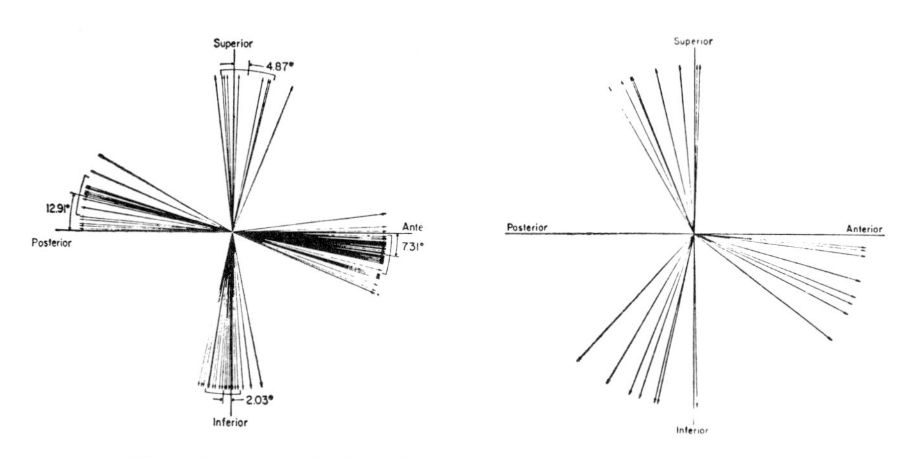

Figure 5 For a sample of on–off type directionally selective ganglion cells the preferred directions of movement fall into four groups (left half), whereas for a sample of the on-type they fall into three groups (right half). The samples came from a fairly restricted region of the visual field, so the explanation offered in Figure 6 is probably correct.

upward movement. In the rabbit there are four types of such neuron with four different trigger features, upward, downward, forward, and backward movement (see Figure 5 above). Since their receptive fields are scattered all over the visual field the rabbit's brain is kept well-informed about movements all around it.

Projective zone Quite unexpectedly, we found that there was another type of ganglion cell which we called the *on-type directionally selective* because, when plotting their receptive fields with a stationary spot, they only responded at onset, unlike the other type I have just illustrated which responded at both onset and offset; we still do not understand the reason for this, but it led us to discover other differences. First, they responded to much slower movements than the other group – so slow, in fact, that they could easily detect the motion of the sun or stars through the sky. But the most surprising difference was that their axes of preferential response were clearly not lined up with the axes of the on–off directionally selective type, and when Clyde Oyster and I analysed this in more detail[2] we found what is shown in Figure 5.

We were completely baffled as to why there should be these differences, and in particular why the preferred directions of the on-type directionally selective ganglion cells should fall into three groups. But I think the answer was discovered several years later.

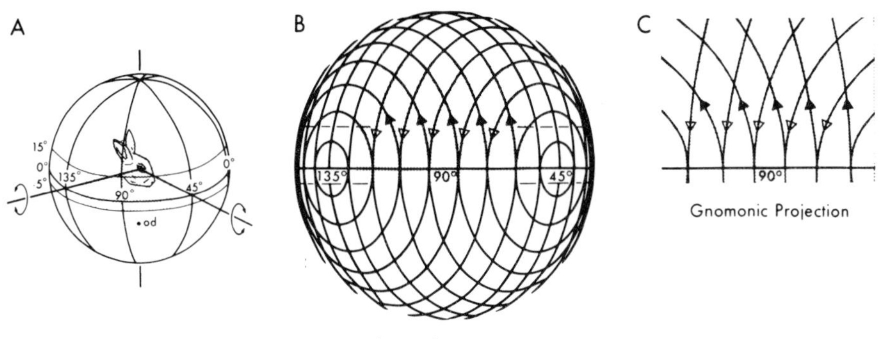

Figure 6 According to J. J. Simpson the on-type directionally selective units signal rotations around three axes which correspond to those around which the three semi-circular canals respond to angular accelerations. As shown in C, this could lead to the appearance of three groups (see Figure 5 from reference 3).

J. J. Simpson[3] and his colleagues showed that the axons of the on-type directionally selective ganglion cells pass to the brain in a special pathway called the Accessory Optic Tract, and end in a nucleus that has three divisions. He suggests that their role is to signal *rotations* of the animal, and to do so using a co-ordinate system that is compatible with the other main organ for signalling rotations of the head, namely the semicircular canals, which are sensitive to angular accelerations. There are three pairs of semicircular canals, which signal rotations around three axes at right angles to each other. Figure 6 shows a rabbit on the left, with these axes marked, two horizontal and the third vertical. The middle diagram shows how objects would move through the visual field for pure rotations round the two horizontal axes, and the right-hand diagram shows, enlarged, the approximate region where the receptive fields shown in Figure 5 had been collected. His hypothesis explains very nicely why we found the three groups, and he has confirmed it by showing that the preferred axes change with position in the visual field in the manner expected.

But in addition to explaining this mystery, it is a beautiful example of what can be achieved by neurons having the appropriate *projective zones*. All the 1,000 or so retinal ganglion cells that are maximally excited by rotation about one particular axis, whatever their positions in the visual field, project to one of the three divisions of the nucleus of the accessory optic tract; similarly all those excited by rotations about the other two also go to their appropriate subdivisions. These arrangements create three small zones in the brain in each of which there will be massive activity for rotation about one particular axis. Collecting together just the right information from scattered origins is obviously an important step in making a sensitive visual analysis of rotation relative to the environment.

Visual information from neurons in these zones supplements that provided from the three semicircular canals and is distributed in parallel with it to the part of the brain that controls our balance, the cerebellum. Incidentally, the ability to respond visually to very slow movements fits in well, for the semi-circular canals fail to respond to slow rotations, so the visual mechanism takes over just when it is most needed, when the inertial mechanism fails.

NEOCORTEX

I hope I have given some idea of what a single neuron with its trigger feature and projective zone can accomplish, and now I must explain a bit about the neocortex, which is the stage on which neurons perform to bring about higher mental processes. First, its origin.

Evolutionary origin About 70 million years ago the brain of the most advanced animals had cerebral hemispheres, but they composed a third or less of the whole brain. They were concerned almost entirely with smell, except for small patches that had inputs from other sensory modalities, probably mainly touch. Presumably this made it possible to combine touch and smell information, though we can only guess about this.

Over the next few million years there was an enormous expansion of this non-smell part of the cerebral hemispheres – the *neopallial explosion* as Elliot Smith[4] called it. He suggested that this occurred because some of the early mammals took to the trees, and in that new environment smell became less important, while vision and neuromuscular control, and particularly the co-ordination of the two, became more important. Similarly, hearing increased in importance both for the information it gave about the environment in general, and for communication within the species; hence the stage was also set for language to be born.

By the time our forbears were established tree-dwellers, the cerebral hemispheres had almost lost their original association with smell, and with the great expansion of the parts devoted to vision, hearing, touch and movement, you have something like the modern primate brain. From then on the most prominent change leading to the human brain was simply an increase in the area of the sheet of cells which forms the neocortex or neopallium. In us, each side has an area of about one square foot, so the crinkled appearance results from the need to crumple the sheet like a piece of paper in order to accommodate it within the skull. But though our brains are big, they are not the biggest; whale and elephant are larger and we only top the scale if you take into account overall body size – our brain is largest relative to our bodies.

Why a structure that had originally been a smell-brain should prove so useful when taken over by other senses is not clear. One possibility is that some superior form of modifiability occurred in the smell-brain, and I shall return to this later. I also like another idea, that because the position of an excited smell receptor is unimportant, its central processing station is adapted to make global rather than local associations. In contrast, the position of an excited tactile, visual, or auditory receptor is all-important and the commonest patterns among them are spatial. One of the features of the cerebral hemispheres is that they have more extensive interconnections from one part to other parts than is usual in sensory centres, and perhaps this was the feature, inherited from its origin as the smell-brain, that made it useful for other modalities as well; global pattern recognition requires taking into account large chunks of sensory information, not just localised patches.

A new creation myth? What I have told you is a 50-year-old version of the creation myth. In some ways it is an improvement over yet older versions, sions, but I hope you realise that the information it encapsulates is much vaguer and less certain than the description I gave of what single cells can do. I suspect it may soon be superseded by a very different idea derived from evolutionary theory and the knowledge molecular biology is giving us about the genetic control of brain processes. Perhaps, for reasons we do not understand, the patterns of interconnections in the smell-brain are under a more versatile system of genetic control, and this allowed the rapid evolution of the neocortex, facilitating the neopallial explosion. It is said that 60 per cent of the human genome codes for proteins that are only expressed in the nervous system, so the real reason that human brains are different may be because greater genetic control has allowed them to evolve more rapidly; if that is the case, taking to the trees may have freed the smell-brain, but it was its susceptibility to rapid evolution that led to the neopallial explosion.

Outline of neocortical structure That is, or may be, how we acquired our brain, but what is its structure? We need to have some idea what the cells are picking up from, and where they are sending information to, in order to apply those two principles I gave you earlier, the selective trigger feature, and the selective projective zone. A converted physicist not afraid of over-simplification, Valentino Braitenberg, has proposed a scheme so simple that it needs no diagram. As pointed out above, it is known that extensive interconnection is a characteristic of the neocortex: thus the great majority of the input to any cell comes from other parts of the neocortex itself, and similarly most of the outputs go to other parts of the neocortex. Braitenberg says these internal connections are of two kinds: *local* interconnections of about 1 mm, which are largely derived from collaterals of axons leaving one region of neocortex and carrying messages elsewhere; and *long-range*, distant connections carried by the axons I have just mentioned. He calculates that there are enough long-range connections for there to be two-way connections between any patch of 1 mm^2 and every other 1 mm^2 patch over the whole neocortex, which is quite impressive when you realise that the human neocortex has an area of more than 100,000 sq. mm.

This pattern of interconnection means that all neurons in the neocortex could connect with any 1 mm^2 region through only one intermediate neuron, and likewise it could be influenced by the activity of any 1 mm^2 region through only one intermediary neuron. Braitenberg's scheme is obviously a great over-simplification which neglects the special connections that exist between regions with related functions, but it may nonetheless point to the overall pattern through which the neocortex carries on its extensive conversations with itself.

Neocortical trigger features After this build-up of the neocortex as the organ endowing us with superior mental functions, it is time to see what messages are actually passed along its nerve fibres. Our knowledge of this comes very largely from David Hubel and Torsten Wiesel, who worked together for twenty years at Harvard Medical School, recording from single neurons of the visual cortex – the part of the neocortex that receives its input mainly from the eyes.[5]

In one of their experiments they first find the region of the visual field that causes electrical activity to be recorded through an electrode placed in a particular patch of neocortex, and then they map the receptive field for a single cell, as illustrated for a retinal ganglion cell in Figure 4. Based on the shape of this receptive field, they then search for the stimulus that gives the best

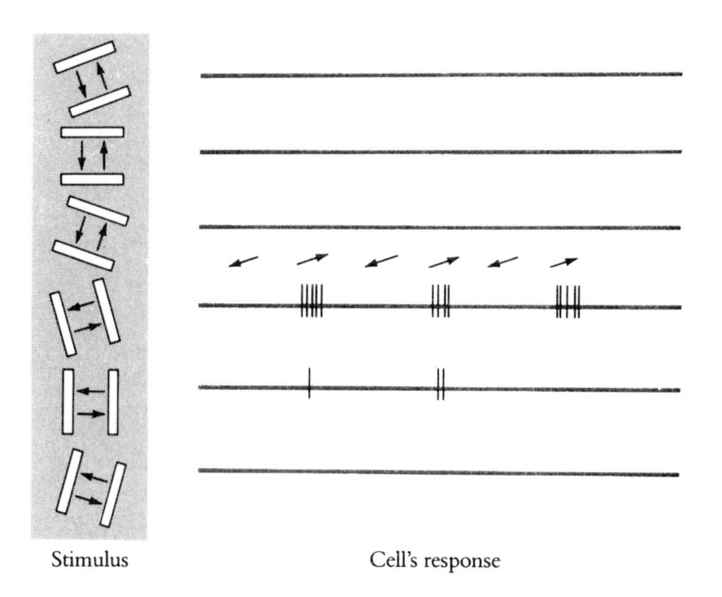

Figure 7 Orientational and directional selectivity in a neuron from the visual cortex. When a white bar is moved to and fro as shown to the left, the neuron responds only when it is nearly vertical, and only for one direction of movement (from Hubel 1988: see Further Reading).

response from it, and in Figure 7 you see that one particular cell responded best to movement of a bar oriented at a particular angle, and also that it only responded when this bar moved in one direction. It turns out that all neurons of the primary visual cortex respond best to oriented bars or edges, though they still vary greatly among themselves as to the position their receptive field occupies in the visual field, as to the direction of preferred orientation, velocity of motion, size of bar and its polarity (dark or light), and in other ways.

The discovery that oriented bars and edges were the trigger features for neurons in the visual cortex was enormously exciting when it was made thirty years ago, but Levick[6] followed up their discovery and found that there are orientation selective neurons in rabbit retina as well as the directionally selective ones shown in Figure 4. Figure 8 illustrates one of these ganglion cells. It had a small receptive field that was somewhat elongated vertically, and on examination with elongated stimuli it became clear that it responded well to a vertical bar but not at all to a horizontal bar. Other similar neurons respond preferentially to horizontal rather than vertical bars, and there are many such ganglion cells to be found in the part of the retina in the rabbit called the visual streak, which normally receives the image of the region of the visual field lying close to the horizon.

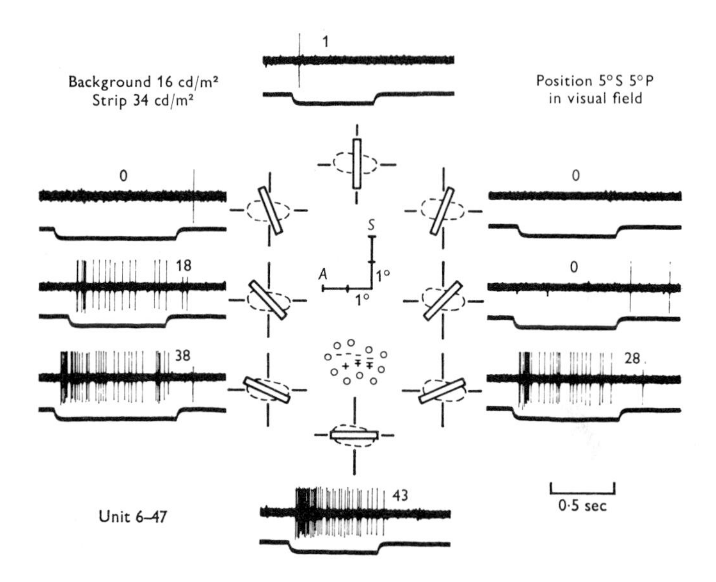

Figure 8 Selectivity for orientation is also found in retinal ganglion cells. This one was recorded from the rabbit by W. R. Levick, reference 6; it had the receptive field plotted in the centre, and responded to bars at different orientations as shown by the responses round the outside of the figure.

There are of course important differences between rabbit retina and visual cortex. The neocortex contains vastly more cells, they come with many different orientational preferences instead of just two, and also in many different sizes. But they are not better pattern selective elements, so I think one can conclude that the orientational selectivity of cortical neurons is not what makes it the organ of civilisation.

Projective zones and non-topographic maps If their special property is not to be found in their trigger features, might it be something about their projective zones? My own view is that the neuron's ability to relay selective information to selected zones has greater possibilities than have yet been discovered by neurophysiologists. Computer scientists appreciate the power of what are called 'generalised Hough transforms', but so far only weak hints of such operations have been found neurophysiologically. The primary visual cortex is surrounded by many secondary visual regions that seem to deal with special qualities of the image such as colour and movement, and in theory much could be achieved by assembling in appropriate ways information about the occurrence of different trigger features.

Perhaps in the next decade the projective zone will be recognised to be a neural processing tool as powerful as the neuron's receptive field and trigger feature.

Already we have a few examples of the end-product of several stages of processing, although we do not yet understand the mechanisms. Many years ago Gross [7] and his colleagues found neurons at higher levels in the visual pathway in which the vigour of the response varied strongly with details of the shape of an object moved about in the visual field. For some cells they had the strong impression that the response increased the more the object resembled a monkey's hand, while in others the trigger feature appeared to be a monkey's face or head. This has been confirmed by others, and cells have even been found that respond preferentially to the face of a particular individual among those that regularly looked after the monkey.[8] Such cells were able to maintain their preference when the heads were inverted or other difficulties placed in the way of recognition, and one would much like to know about the computational principles and physiological mechanisms that performed the task.

In spite of these impressive results, progress in finding other examples of neurons with highly selective trigger features has not been as rapid as was hoped. Most people feel there must be other principles at work to explain high level functions, and the plasticity of neurons is an obvious place to look.

MODIFIABILITY AND LEARNING

Actors have to be trained, and they have to learn their parts; similarly, neurons must somehow acquire their trigger features and projective zones. It is easy to see that studying the growth and modifiability of neurons is a much harder task than describing them in the state you normally find them, so it is not surprising that much less is known, and all I can do here is to point out some of the interesting possibilities that are opening up.

There appear to be two types of modifiability in cortical neurons: a *slow, irreversible* process that occurs during early development, and a *rapid, reversible* one that continues to operate through life.

Rapid, reversible modification This rapid process is responsible for what psychologists call figural after-effects, contingent after-effects, or pattern specific adaptation; Figure 9 demonstrates two forms of it. First, look at the mark between the central pair, and note that top and bottom are the

Figure 9 Short-term effects of experience. Gaze at the black bar between the left-hand pair of slanting gratings for about thirty seconds, then transfer your gaze to the dot between the central pair of gratings; they will appear to slant in the opposite directions. The right pair of gratings give a corresponding after-effect for size (from Blakemore, reference 13).

same size and are both vertical. Next, look at the line between the gratings on the left for about thirty seconds, then transfer your gaze to the central pair again. They no longer look vertical, but slant in opposite directions. Do the same for the right pair, and the central pair will look unequal in size. There are many other simple demonstrations of similar phenomena, and the oldest of them, the after-effect of movement, was known to Aristotle.

At a phenomenal level, what happens here can be described by saying that it is as if your perceptual mechanisms became fatigued by some salient characteristic of the adapting stimulus – its orientation or periodicity in the case of Figure 9, or direction of movement in the case of the motion after-effect. The popularity of this explanation gained a lot from the demonstration that neurons in sensory pathways respond selectively to such characteristics, and do indeed show a decline in their response on continued exposure. But recently my colleagues and I have developed another theory which I think has more interesting implications.[9] The strengths of the theory are, first that it explains the perceptual phenomena I have just described, second that the brain has a real need for the operation it postulates, and third that it gives a role to a prominent but hitherto unexplained anatomical characteristic of the neocortex. I will start with the prominent anatomical fact.

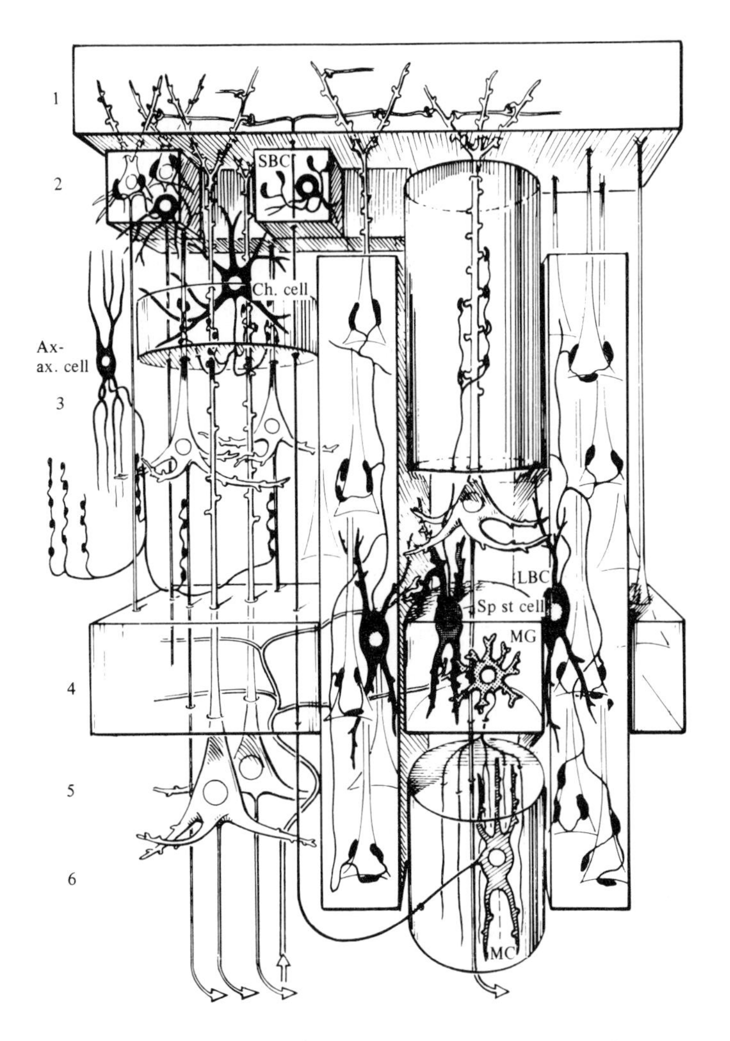

Figure 10 Ten years ago the famous Hungarian neuro-anatomist Janos Szentagothai drew this picture of a section of cortex about 1 mm across. For a modern view, which is no simpler, see the review by Kevan Martin, reference 10.

Positive feed-back Figure 10 shows what Janos Szentagothai, a famous Hungarian neuro-anatomist, conceived the neocortex to be like ten years ago.[10] It can be described briefly by a single word – complicated. And the diagram is much simpler than the real neocortex, which contains 10,000 cells where only 25 are shown here. But one simple fact emerges, namely that there are innumerable interconnections between the cells in one small region, which, of course, only reinforces what I have said several times already, namely that the neocortex talks a lot to itself. In what follows I shall describe one type of such interconnection, though I confess I am not by any means sure that it is the one responsible for the rapid modification process.

It is possible to find the origin of the synapses ending on cells of the type Hubel and Wiesel recorded from, and it is an amazing fact that 80 per cent or more of their excitatory input comes from other cortical cells, and less than 20 per cent from the nerve fibres coming up from the eye. One would expect this large amount of positive feed-back to lead to instability: as soon as the input is strong enough to excite one or two neurons, one would expect these to feed back on to others and initiate an explosive chain reaction. But suppose there is a rule which says that the effectiveness of a synapse decreases whenever it causes the post-synaptic cell to fire. This will stabilise the network, for the amount of positive feed-back between those combinations of neighbouring cortical neurons that tend to be active together will rapidly decline.

After a time the circuit should settle down to a state in which the *usual* combinations of inputs to the cell are relatively ineffective, while any *unusual* combination can still set off a strong response. Thus the combination of positive feed-back with the rule defined above has the making of a device for detecting unusual combinations. The detection of novel conjunctions, or suspicious coincidences, is the essence of good detective work, so this simple mechanism might endow our actors, the nerve cells, with a grain of intelligence.

Need Networks operating on this principle perform an operation that is likely to be extremely important for the neocortex, and it was actually the search for a mechanism that would do this that led us to the suggested modification rule: the modifiable interconnections tend to make the representative elements become uncorrelated, and thus to signal independently of each other. It is obvious enough that a neocortex would be useless if all its neurons responded to the same or strongly similar features, so *decorrelation* is useful for this purpose, but it also confers another great advantage. If the responses of a set of representative elements are statistically independent of each other, it becomes relatively easy to form reliable associations with combinations of them, whereas this is much more difficult when the responses show strong correlations. Thus decorrelation would enormously increase the versatility of the cortex in detecting new associations. Increased versatility of learning would confer great selective advantage and is just the kind of

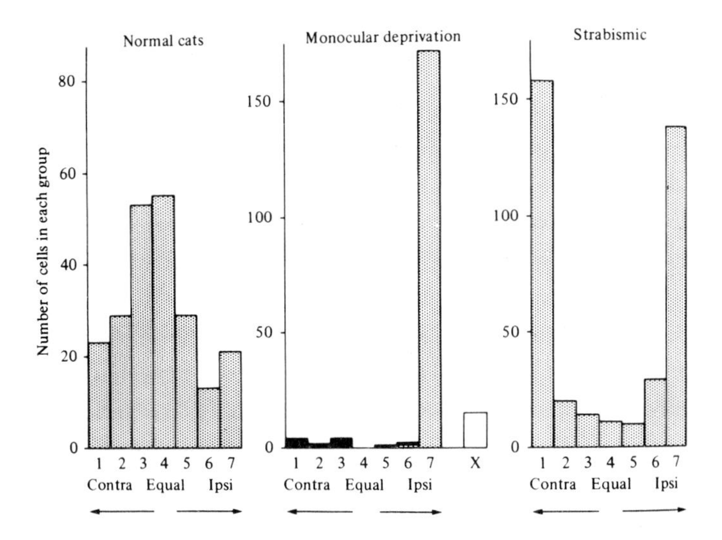

Figure 11 Long-term effects of deprivation during the sensitive period. The histograms show the numbers of cells controlled exclusively by the contralateral eye (1) or ipsilateral eye (7), with intermediate balance of control shown in groups 2 to 6. Normal cats have the distribution shown at the left, cats deprived of vision in the contralateral eye during the first three months of life have the distribution shown in the middle, and cats whose eyes were misaligned as a result of surgery have the distribution shown at the right. Experience is required for the connections from an eye to continue to control its fair share of cortical neurons, and joint experience is required for joint control (adapted from Hubel and Wiesel, reference 5 and Hubel 1988: see Further Reading).

change that might lead to the very rapid evolution of neocortex, so we can add this to global connectivity and greater genetic control as possible causes of the neopallial explosion.

This is all I can tell you about the rapid modification process; it has the effect of diminishing the prominence of events and combinations of events that normally occur together, giving prominence to new associations. The slow modification process is quite different.

Slow modification Hubel and Wiesel[11] showed that the properties of the neurons in the visual cortex were drastically changed by restricting the visual experience of an animal, provided that these restrictions took place during a few months early in life called the sensitive period. Figure 11 shows the result of one of their experiments. They were interested in the way that the connections between an eye and a cortical neuron were affected by whether that eye had or had not been used, so first they measured the numbers of cells establishing these connections in normal animals. These numbers are given in the histogram at the left, 1 meaning that a cell could only be activated through the contralateral eye, 7 that it could only be activated through the ipsilateral eye, and intermediate numbers indicating intermediate degrees of ocular dominance; thus group 4 means that both eyes affected the neuron equally. They then reared some animals with one eye closed throughout the sensitive period, and as you see the result was dramatic – the closed eye lost control of almost all the neurons it would normally have activated.

Carrying the experiment a stage further, they tried the effect of misaligning the eyes. Both eyes now had the same average experience, but since they were misaligned, cells would no longer be activated simultaneously by both eyes. The left histogram shows the result: each eye established control over roughly equal numbers of neurons, but the number connected to both was dramatically less than in normal animals. Apparently, simultaneous activation of a neuron through both eyes is required to maintain connections to both eyes.

Similar experiments have been done by Blakemore, Hirsch and Spinelli, and others, which show that the same rule applies to other forms of selectivity.[12] For instance, if vertical lines and edges have not been experienced, there will be few cells responding to vertical lines and edges. One wonders how many of one's own mental defects result from lack of the appropriate experience at the proper time!

Notice that, at a phenomenological level, the slow modification process works in the opposite direction to the fast modification process, which desensitised the system to the adapting stimulus. In the slow process it appears that exercise of a particular type strengthens, or at least preserves, the system's response to that type of stimulus. It is particularly interesting that this strengthening or preservation applies to the conjunction of two stimuli, as shown in the strabismus experiment: cells that respond to joint excitation of the two eyes are normally found, but are missing if joint excitation has been made very improbable by misaligning the eyes.

It has been argued that the mechanism of synaptic reinforcement at work here follows a rule that the psychologist Donald Hebb proposed forty years ago as the neuronal basis of memory: when an input pattern is successful in making a neuron fire, then the synapses that took part in making it fire are strengthened. These modifications are relatively slow, probably requiring hours or days, and they are thought to depend heavily on growth processes.

The rule for rapid modification suggested earlier was actually anti-Hebbian, the opposite to the above since it postulated a decrease of synaptic efficacy whenever presynaptic activity successfully excited the post-synaptic neuron. But rapid anti-Hebbian modification of the mutual interactions between the neurons at one level should work rather well if combined with slower Hebbian modification of the synapses feeding excitation from a lower level. The fast process would ensure that each cortical neuron tended to respond to an input feature that was different from the features of other cortical neurons. Then, since it is whether or not a cell actually responds that determines whether it undergoes the slow modification process, the slow process would make cortical neurons become permanently tuned to this uncorrelated set of features. Thus the fast process may make a permanent contribution during the sensitive period, while its influence later would adjust the cortical feature detectors, tending to make them respond most to the unusual combinations and coincidences in their input.

CONCLUSIONS

Now I have explained all I can, you will realise that we really do not understand why the neocortex is, as Herrick called it, the organ of civilisation. We need not doubt that the epithet is justified and I think we are on the right track in attributing its powers to nerve cells with their trigger features and projective zones, but the connecting links are missing. Thus our next problems are to understand what principles govern the establishment and modification of the interconnections between unimaginable numbers of nerve cells, how these interconnections endow us with higher mental functions, and the cellular mechanisms that bring about these remarkable processes. Perhaps we are just beginning to see how the brain works, but we have a long way to go.

FURTHER READING

Braitenberg, V. *On the Texture of Brains*. New York: Springer Verlag, 1977.
Herrick, C. J. *Brains of Rats and Men*. University of Chicago Press, 1926.
Hubel, D. H. *Eye, Brain and Vision*. Scientific American Library series, no. 22. New York: distributed by W. H. Freeman, 1988.

REFERENCES

1 H. B. Barlow, R. M. Hill and W. R. Levick, 'Retinal ganglion cells responding selectively to direction and speed of motion in the rabbit', *Journal of Physiology*, 173 (1964), 377–407.
2 C. W. Oyster and H. B. Barlow, 'Direction-selective units in rabbit retina: distribution of preferred directions', *Science*, 155 (1967), 841–2.
3 J. J. Simpson, 'The accessory optic system', *Annual Review of Neuroscience*, 7 (1984), 13–41.
4 G. Elliot Smith, *The Evolution of Man*, Oxford University Press, 1924.
5 D. H. Hubel and T. N. Wiesel, 'Functional architecture of macaque monkey visual cortex' (The Ferrier Lecture), *Proceedings of the Royal Society*, B 198 (1977), 1–59.
6 W. R. Levick, 'Receptive fields and trigger features of ganglion cells in the visual streak of the rabbit's retina', *Journal of Physiology*, 188 (1967), 285–307.
7 C. G. Gross, 'Inferotemporal cortex and vision', *Progress in Physiological Psychology*, 5 (1973), 77–123.
8 D. I. Perrett, P. A. J. Smith, D. D. Potter, A. J. Mistlin, A. S. Head, A. D. Milner and M. A. Jeeves, 'Neurones responsive to faces in the temporal cortex: studies of functional organisation, sensitivity and relation to perception', *Human Neurobiology*, 3 (1984), 197–208.
9 H. B. Barlow and P. Földiák, 'Adaptation and decorrelation in the cortex' in R. Durbin, C. Miall and G. Mitchison (eds.), *The Computing Neuron*, Wokingham, England: Addison Wesley, 1989, pp 54–72.
10 K. A. C. Martin, 'From single cells to simple circuits in the cerebral cortex' (The Wellcome Lecture), *Quarterly Journal of Experimental Physiology*, 73 (1988), 637–702.
11 Hubel and Wiesel, 'Functional architecture of macaque monkey visual cortex'.
12 For a review see J. A. Movshon and R. C. Van Sluyters, 'Visual neural development', *Annual Review of Psychology*, 32 (1981), 477–522.
13 C. Blakemore, 'The baffled brain' in R. L. Gregory and E. H. Gombrich (eds.), *Illusion in Nature and Art*, London: Duckworth, 1973, pp. 8–47.

2

Animal communication

PATRICK BATESON

Equipped with his ring, King Solomon is commonly supposed to have spoken *with* animals. The biblical text suggests simply that he spoke *of* them, knowledgeable man that he was. I suspect the misinterpretation represents the deep longing that most humans have to get inside the heads of other animals. If anybody could do it, surely Solomon could. Of course the fantasy of animals talking like humans is as old as human fables and is resplendent in a great chunk of the literature for children. (Not least, Winnie the Pooh, pondering on truth in chapter 4.) We have a wonderful capacity, some would say an over-developed capacity, to project ourselves into other beings. If we are imaginative enough we can project ourselves inside plants and inanimate objects as well as other animals. We wonder what it would be like to be an oak tree, a house, a mountain, even a thunder cloud.

The urge to empathise is strong, but the projection is often rewarded by understanding in the case of animals. People who know their animals well have a strong sense of what they are going to do next from what is referred to as the animals' 'body language'. Certainly, pet owners and expert ethologists alike can generally tell from cats' and dogs' expressions whether they are likely to attack or escape. We sense that their postures represent mixtures of the human emotions of fear and aggression. It may seem reasonable, but we should be clear that the arguments used in discussions about intention, emotion, pain and language in other animals are usually extrapolations from ourselves. We have intentions, feel pain, and tell each other things. If we believe that humans have evolved, we are liable to assume that our subjective experiences are shared by other animals. This evolutionary argument appeals to continuities between humans and other animals, looked at from a human point of view.

Another utterly different approach to animal communication involves asking what the observed activities might be for. In other words, what is the evolved function of these behaviour patterns? This evolutionary argument asks what it is about a feature that improves the animal's chances of surviving and reproducing itself. If we consider why communication might have evolved, we are drawn into types of arguments and speculations which are much less familiar to the lay-person.

DARWINIAN EXPLANATION

Virtually every biologist who cares to think about the subject believes that all living matter has evolved. Existing species were not created in their present form at the beginning of life on this planet. The modern scientific debates are about *how* the changes came about, not about whether or not they happened. Chance and catastrophe are unsatisfying and inadequate as explanations when we try to understand the numerous and exquisite examples of correlations between the characters of organisms and their physical and social environments. Such adaptations grab our attention because the characters seem so well designed for the job they perform. Much the most coherent explanation for the evolution of such phenomena is still Charles Darwin's. Indeed, Darwin's proposal is much better seen as a theory about the origin of adaptations than as a theory about the origin of species.

Darwin's proposed mechanism depends crucially on two starting conditions. First, variation in a character must exist at the outset of the evolutionary process. Second, close relatives must resemble each other with respect to such a character more than do distantly related individuals. The steps in the process involve some individuals surviving or breeding more readily than others. If the ones that survive or breed more easily carry a particular version of the character, the character will be more strongly represented in future generations. If the character enabled them to survive or breed more readily, then the long-term consequence is that the character will generally bear some relation to the conditions in which it worked.

Richard Dawkins[1] has argued that individual organisms do not survive from one generation to the next, while on the whole their genes do. He proposed that, therefore, Darwinian evolution is primarily about changes in the genes. Dawkins's approach to evolution was presented in characteristically

Figure 1 A reed warbler feeding a young cuckoo as if it were its own offspring (Photo: I. Wyllie).

entertaining form when he suggested that the organism is '... a robot vehicle blindly programmed to preserve its selfish genes'. His approach has undoubtedly helped a lot of people to understand the complexities of biological evolution. While Dawkins cannot be blamed for it, modern enthusiasm for the 'enterprise society' may explain why his parable of selfish genes has commonly been elided with the selfish intentions of individuals. Be that as it may, the general style of thinking about evolution has been applied to the study of animal communication in ways that suggest that all activities directed by one individual towards another are manipulative. Sometimes this view is clearly correct. We see it in its most obvious form in the interactions between species when one species manages to control the behaviour of another as if it were a puppet. A striking example is the European cuckoo. The mother cuckoo lays each egg in the nest of another species such as the reed warbler. The egg very closely resembles the egg of the host. The young cuckoo hatches before the reed warblers and ejects the competition from the nest. Then the young cuckoo successfully persuades the unfortunate warbler parents to feed it, even when it is twice their size (Figure 1). By looking like a super-offspring, the cuckoo successfully exploits the normal pattern of interaction that exists between parent and young. Is it the case that *all* communication within a species should be treated as though it were part of selfish manipulation? The conclusion seems most sensible when applied to the displays between rivals for a mate, for food or for territory.

SIGNALS IN CONFLICTS

At the moment there is vigorous theoretical argument among people who study animal signals, with some taking the extreme position that signals carry no accurate information about the state of the signaller and others arguing that the signals are extremely reliable. Those who advocate total manipulation urge that a trustworthy mode of communication is always open to cheating.[2] The cheat uses the information provided by its opponent and gives nothing away about itself. As a consequence, it is much more likely to win the resource. And, as a result of that, it is likely to reproduce faster. Before long, cheating will have evolved to become the dominant mode of behaviour.

It sounds convincing but, as I have already noted, most people who know a particular species well quickly develop a good intuitive sense from an animal's bodily and facial postures of whether it is likely to attack or escape. Examples from the domestic cat and the wolf, which is the ancestor of the domestic dog, are shown in Figure 2. Most pet owners who know either cats or dogs would readily agree with where the expressions have been placed on the two axes of likelihood of attack and likelihood of escape. A great deal is now known about such signals in a wide range of animals, from crabs and spiders to monkeys and humans. The theories about how the signalling behaviour is controlled are also developing fast. As an example, take a spider in which each individual needs to find a good place in which to put its web, but good places for making webs are in short supply.[3] Contests over web-sites consist of a series of bouts during which the spiders are increasingly

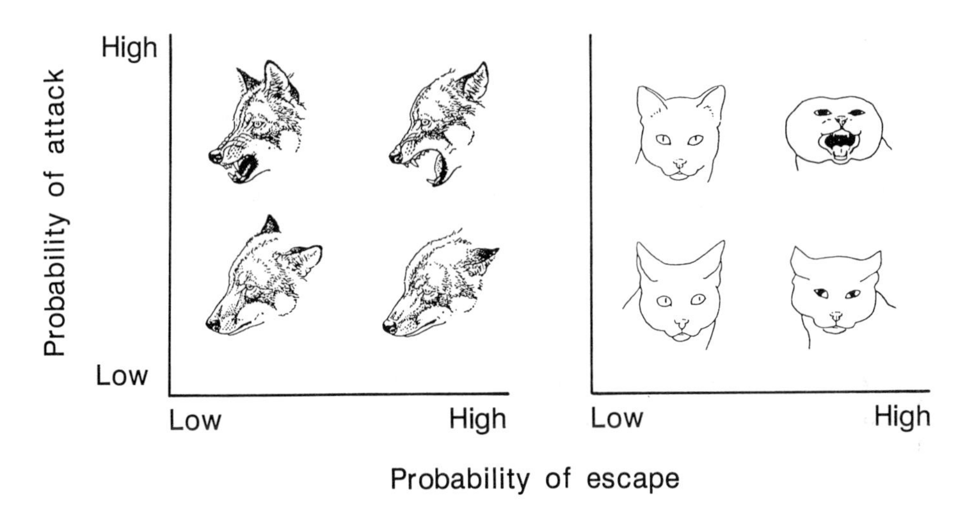

Figure 2 Expressions given by the domestic cat and the wolf in different combinations of circumstances (from drawings by P. Leyhausen of the cat and P. Barrett of the wolf).

likely to attack each other if the contest persists. After the initial location of the opponent, three levels of escalation precede an actual fight, involving easily recognised and increasingly energetic displays. Neither party will benefit from getting hurt and in the great majority of cases the disputes are settled without serious damage on either side. The encounter can break off at any stage in the process of escalation. The contests are usually won by the large spider, but if the spiders are of equal size, they are usually won by the owner. The displays are usually more energetic if the site is a particularly good one for snaring prey. Even in these spiders, the simplest explanations for what they do requires that they assess their own fighting ability relative to that of their opponent. They also need to assess their own needs and how much their opponent is prepared to fight. The theoretical model, developed by Maynard Smith and Reichert[4] to explain what happens, also requires that the behaviour of each individual at each stage of escalation indicates how serious it is about continuing.

All this seems to contradict the expectation of the manipulation theorists. If we can tell what the animals are about to do next, then surely so can their opponents. It is sometimes claimed, however, that in fact we are fooling ourselves and our capacity to predict what will happen is not very good.[5] From a large number of quantititative studies, it is clear that escape *is* very well predicted by certain patterns of behaviour. For instance, two postures

Figure 3 Three blue tits threatening each other at the bird table (Photo: Kim Taylor/Bruce Coleman Ltd).

given by blue tits threatening each other at a bird table accurately predict escape about 90 per cent of the time (Figure 3). We can see some sense in this. If one animal suddenly turns tail, it is liable to be attacked and might get injured. The advantage to the loser of not being misunderstood and expressing the animal equivalent of a white flag are obvious. The winner's benefit from responding appropriately to such a signal is that it does not risk injury by escalating the conflict into a real fight when it need not have done so. The argument is, therefore, that a form of behaviour which effectively negotiates the end of a conflict can be evolutionarily stable.

That having been said, it is true that attack is predicted by other postures only about half as well as escape.[6] This has been taken as evidence for the manipulative view of animal communication seen in conflicts. My guess is that you would draw a comparable, but equally misleading, conclusion from similar analysis of the human game of chess where cheating is not possible. Tipping the king over in resignation tells you without fail who is the winner. On the other hand, it can be extremely difficult to be sure who is going to win until very near the end of the game. In a famous encounter between Robert Byrne and Bobby Fischer in 1963, Fischer seemed to have lost the game by the 21st move. He had just moved his Queen and two grandmasters, providing a commentary for spectators, declared that Byrne had a won game. In fact,

Byrne knew better and eventually resigned without making another move because he realised that he was due to be mated within four moves! Because of the variety of possibilities at each move, the ability to predict who will win from a given move becomes less and less reliable, the longer the game has to run after that move. Therefore, the empirical evidence suggesting that reliable signals of each actor's state are not seen in the course of animal conflicts may have been misinterpreted. These signals may have been less predictive of the outcome because they occurred at higher frequencies at earlier stages in the conflicts.

Even so, the issue is complicated. The crucial question for evolutionary biology is where the balance is struck between signalling real information about your state and signalling misinformation. Consider analogies with another human game, poker. On the one hand, if you can get away with bluff, you make a lot of money. On the other, if you bluff against an opponent who has a really good hand you may end up very much worse off than if you had decided to throw in your bad hand before you had raised the bet too far. So we might well expect something equivalent to negotiation, as Robert Hinde[7] has argued. For each individual the optimal outcome of such negotiation should represent a balance between the costs of escalating the conflict to likely injury and the benefits of winning the resource easily. An individual that escalates without assessment is in danger of finding itself in a fight with a much stronger individual. This is not purely a verbal argument since recently Alan Bond[8] has shown how an equilibrium between deception and honesty might be struck in the course of evolution.

The developments in thinking about what happens in animal conflicts have some interesting parallels with economists' theorising about human bargaining. For many years it was supposed that nothing should be revealed about the bargainer's intentions.[9] What the opponent observes is either an attempt at manipulation or costly delays by the other party, expressing apparent lack of interest. However, subtle signals, known as 'cheap talk' are now recognised.[10] These signals may maintain negotiations when they might otherwise have broken down and, therefore, benefit both parties by increasing the likelihood that a transaction ultimately takes place. This is somewhat similar to what is now recognised in animal conflict.

The evolutionary pressure for some exaggeration of fighting ability is present and many species puff themselves up in various ways, thereby making themselves look more fearsome than they really are. Even so, honest advertisement of strength providing cues that cannot be faked may count

Figure 4 A red deer stag roaring during the mating season (Photo: T. Clutton-Brock).

most in the long run. While fake characteristics may gain short-term successes, those individuals that ignore such clues and focus on reliable sources of evidence about their opponents will eventually emerge in the course of evolution. One example may be the male red deer which competes vigorously with other males for opportunities to mate. Fights occur, of course, but conflicts are most often settled by bouts of roaring at each other (Figure 4). The rate of roaring is increased by each individual as the contest proceeds until one individual gives up. The roaring is extremely exhausting for the animals and the one that can keep it up for longest is also likely to be the strongest. In the roaring contests, both individuals increase the rate of roaring until one seems to recognise it is outclassed and retreats. Tim Clutton-Brock and Steve Albon[11] played tape-recordings of red deer roars to a real stag. Unlike the stag, the tape-recorder did not get tired and when the tape-recorder roared at high rate, the stag roared less. It seems as though the stag has been forced into accepting that it was dealing with a much more powerful opponent.

Figure 5 The courtship displays of the male mandarin drake in which he shows off his striking orange wing feather (from *Owen: Wildfowl of Europe*, by kind permission of Macmillan, London and Basingstoke).

SIGNALS IN COURTSHIP

I will turn now to a much easier, less controversial case, namely courtship before mating. Being explicit about internal state is an obvious advantage to the signaller as well as to the receiver. If he or she indicates in effect, 'I'm ready to mate', and the signal is correctly interpreted, then the arrangement is advantageous to both sides. In the animal kingdom, we find many such examples of signals that play an important role in courtship. The drake mandarin duck has an enormous orange flag on each wing modified from one of the wing feathers (Figure 5). In courtship the male turns his head and points at this beautiful feather with his beak.

How did the posture and its supporting feather ever evolve? In other ducks courting males may also preen themselves in certain phases of their displays. Ethologists often notice that animals that are about to switch from doing one thing to doing something else, first do something quite irrelevant. These so-called 'displacement activities' which often involve grooming the body for a short time, seem to occur particularly in states of conflict between, say, attacking and escaping from another individual. When two high priority activities such as approaching and moving away are in balance, lower priority activities may be briefly observed. So it is possible that, in the course of evolution, the preening activity of the male has been accentuated because it signals to a female that he is in a state of conflict. Now, in the early stages of courtship, the male is aggressive and if the female comes too close he is likely to attack her. If the male, by his displacement activity, conveys that he is in a state of conflict between attacking the female and behaving sexually towards her, she could detect that this male is interested in her. As a consequence she may be more likely to mate with him than with a male who is less explicit about his state. The individual that produces the most clear-cut signal is most likely to have the most offspring. In this way, an increasingly ritualised signal might have evolved.

It is possible that, if the signal was made even more dramatic by being supported by a set of feathers, the chances of a male getting a mate and thereby breeding might be even further enhanced. However, combinations of characters can produce surprising outcomes in evolution as in everything else. In many species mating preferences are acquired by a learning process known as sexual imprinting. In my own work with Japanese quail I have found that this process may lead to a preference for a partner that is slightly novel – just a bit different but not too different from the members of the opposite sex it knew when it was young (Bateson, *Mate Choice*). If the quail have been reared with siblings, both sexes prefer to mate with first cousins. The reasons for doing this may be that the animal maximises its chances of mating with somebody with whom it can have offspring while at the same time minimising the ill-effects of inbreeding. At the same time, for quite different reasons, animals have evolved sensory systems that are particularly good at picking up biologically important features of the environment. In the visual system of birds, colours and contrasting outlines are some of the features picked out and responded to strongly. When these two separately evolved features came together, some interesting evolutionary changes became possible.[12] Suppose that the mating preference was asymmetrically arranged around the familiar so that first cousins with conspicuous plumage were preferred over those with dowdier plumage, then there would have been a relentless pressure for plumage to become more conspicuous. We cannot, of course, replay a piece of history and we do not have a fossil record of behaviour. But we can at least test the logic in computer simulations. Of course, the pressure for change may be resisted because birds with brightly coloured or greatly elongated feathers are vulnerable to predators. Since lots of birds have highly conspicuous plumage, the possession of such plumage would seem to carry with it some real advantages.

Figure 6 A cleaner fish unharmed inside the mouth of predator fish (Photo: Bill Wood/Bruce Coleman Ltd).

SIGNALS IN CO-OPERATION

The evolutionary explanations I have used are all couched in terms of the benefit for the individual. The same logic can be brought to bear on the cases in which individuals co-operate. Even though Darwinian evolution is represented as a competitive process, the outcome has often been that animals ended up working with each other. One explanation for co-operation is that, at least in the past, the aided individuals were relatives; co-operation is like parental care and has evolved for similar reasons. Another is that co-operating individuals jointly benefited even though they were not related; the co-operative behaviour has evolved because those who did it were more likely to survive as individuals and reproduce than those that did not. Once again the force of this particular argument can be seen most clearly in communication between different species.

Little striped fish clean the teeth of great big predator fish. Before they do their job, the cleaner fish do a characteristic waggling swim in front of the monster. This inhibits the normal feeding response and the great predator opens its mouth, allowing the little fish in (Figure 6). When the big fish needs to eat other little fish, it signals it is switching back into normal hunting mode by jerking its jaw in a particular way. The little fish scuttles for cover and the symbiotic arrangement is preserved. Both parties benefit by this arrangement.

I will give some examples of signals that maintain relationships within a species from a familiar companion animal, the domestic cat. The cat's independence has encouraged a widespread view that it is asocial and unco-operative. However, the studies of cats living under natural conditions have revealed that, apart from an intense early family life, the females in particular may stay in groups as adults.[13] While living together, cats may help each other in terms of mutual defence against intruders and caring for each other's offspring. Of course, the benefits to the individual of co-operation change as conditions change and, in really difficult circumstances, previously existing mutually beneficial arrangements may break down. Or if members of a group are not familiar with each other, no mutual aid may occur until they have been together for some time. As familiarity grows, individuals come to sense each other's reliability. Furthermore, expectation of an indefinite number of future meetings means that deception or conflict are much less attractive options. Once evolutionary stability of co-operative behaviour under some conditions had been reached, features that maintained and enhanced the coherence of the behaviour then evolved. For instance, purring almost certainly signals that the cat is relaxed and contented. Kittens first purr while suckling when they are a few days old. Their purring probably acts like the smile of a human baby indicating to the mother that all is well. If so, the purr helps to establish and maintain a close relationship. Probably for similar reasons, the purr is used by adults in social and sexual contexts. For instance, an adult female will purr while suckling her kittens and when she courts a male. Again like the human smile, purring can be used in appeasement by a subordinate animal towards a dominant one, the implication being that it reduces the likelihood of attack. Cats frequently rub parts of their body against objects and other animals (Figure 7). The patches between the eyes and the ears (which are only lightly covered with fur), the lips, and chin and the tail are all richly supplied with glands producing fatty secretions. The lips, chin and tail are primarily used in marking objects and the head patches and also the tail are used in marking other cats. The result of marking with the head patch may sometimes

Figure 7 A domestic cat rubbing the hand of a human with the glands between the eye and the ear (Photo: L. Barden).

be seen if a friendly cat on the other side of a window can be persuaded to approach and rub. If the light is right, a broad smear, which quickly dries, may be seen where the cat has pushed its head against the glass. Given that other cats are marked with the patch and the rubbing is reciprocated, it would seem that all the cats in a social group end up smelling alike. If that is so, then the common odour would be an olfactory badge which might denote common membership of the club.

One of the most characteristic signals of a cat entering or leaving a social group is the raising of its tail. The raised tail is a visual signal to the others (as it is to humans) that the individual is relaxed and friendly. Such signals may be performed regularly because, like a human hand-shake, the cat maintains stable social relationships in this way and reduces the chances that it will be disrupted in its daily round by the other individuals with which it lives. Finally, another friendly gesture is the blink. A prolonged stare is intimidating and may cause a subordinate cat to withdraw. Perhaps for this reason, non-aggressive cats when staring at other cats or at humans will blink, thereby signalling that the scrutiny is not hostile. In Darwinian terms, once again, cats that did this were more likely to maintain their social relationships and thereby derive the benefits that such relationships provide than cats that did not.

In co-operating animals, the mutual benefits of working together can be greatly enhanced if information about the state of the external world can be transmitted from one individual to another. One of the most extraordinary and well-analysed examples of such transmission is still provided by the so-called dance language of honey bees described so beautifully by Karl von Frisch. The story is well known. The characteristics of the waggle dance performed in the hive provide the crucial information about where the returning bee successfully foraged. The duration of the dance circuit is strongly correlated with the distance from the hive to the food. In a darkened hive the angle of orientation of the central segment of the dance with respect to the vertical is strongly correlated with the angle between the food source and the sun's position. Now, it was logically possible that the cue used by the recruits following a dancer was something other than the angle of the dance – and for a while von Frisch was sharply criticised for precisely this reason. However, an ingenious experiment by Jim Gould[14] has provided direct support for the view that the new recruits use the angle of the central segment of the waggle dance.

While bees will orientate their dances with respect to gravity in a darkened hive, in bright light, they orientate with respect to the light. The switch from one mechanism to the other is done by the stimulation of three light-sensitive cells between the big compound eyes on their heads. If these are painted black, bees orientate their dances with respect to gravity even in bright light. In the experiment, a light of adequate brightness was placed at 85° to the left of the vertical in the hive. The first foragers to a feeding site were captured and their light-sensitive cells were painted so that when they returned to the hive they orientated their dances with respect to gravity. Since the new recruits in the hive had not been tampered with and were stimulated by the light, they misinterpreted the dance of the returning foragers. As a result they flew out from the hive at an angle of 85° to the right of where the food actually was. So it seems virtually certain that the angle of the central part of the waggle dance relative to either gravity or a bright light does indeed provide the crucial information for the new recruits.

In general, it can often be difficult for us to be sure just what is going on in supposed examples of communication. Even so, it is possible to get a long way with well-designed experiments. Another example is provided by African vervet monkeys. These monkeys produce alarm calls when one of their many predators approaches – thereby alerting other members of the group.

They are particularly interesting because they produce several calls and each call is specific to a particular type of predator: a low grunt in response to eagles, a high chutter in response to snakes like pythons and a rather pure tone in response to leopards. Simply observing vervets rush into trees when somebody gives a tonal call does not guarantee that it was the call that did it. The other monkeys might have been responding to a visual signal or they may have seen the leopard themselves. The importance of the calls was demonstrated nicely by Dorothy Cheney and Robert Seyfarth.[15] They first recorded the vervets' call and then played them through loud-speakers to free-living monkeys moving about on the ground. When Cheney and Seyfarth played a tonal call normally given in response to leopards, the majority of the monkeys ran to a tree. When they played a low grunt normally given in response to eagles, the majority of the monkeys looked up. And when they played a high chutter, normally given in response to snakes, the majority of the monkeys looked down.

LEARNING ABOUT SIGNALS

Unlike the honey bees, a great deal of the vervet's signalling system is learned. While the adults distinguish between particular predator species within a class and only call to species that are likely to give them trouble, the young vervets give leopard alarm calls in response to a wide variety of terrestrial mammals, the eagle calls in response to a great many different birds and snake alarm calls in response to long thin objects, many of which are not snakes.

Many birds and mammals may have alarm calls that are given to objects that other members of their species have indicated are frightening. It turns out that like the vervets, birds will also learn to whom they should respond. For instance, blackbirds will mob owls, producing a high intensity pinking call. When a stuffed owl was presented to experienced blackbirds living in aviaries, they started mobbing the mounted bird.[16] If, meanwhile, another blackbird saw a species it has never seen before, it would rapidly associate this species with the mobbing calls made by its fellows. The novel species was an Australian bird that the European blackbird was not likely to have seen before, called the noisy friar bird. This grotesque bird is specialised for taking honey from wild bees' nests and has no feathers on its head. The strangeness and oddity of the stuffed Australian bird were enhanced by painting its head blue. The blackbirds that saw it quickly treated this bird as though it were potentially dangerous and mobbed it, even when their fellows could not see the owl and so were making no noise. In later experiments a brightly coloured plastic bottle was used and the blackbirds even learned to mob this object, although it was not so effective as a novel stuffed bird. Evidently there are some constraints on what the blackbirds will learn to treat as enemies. In many ways this is like our own fear of snakes. We have a predisposition to avoid them, but this can be enormously enhanced by the alarm of others.

This experiment showed how important learning can be in interpreting a signal. Not surprising, you might think, in animals designed by Darwinian evolution to find associations in the real world. But it emphasised the point that the capacity to signal and the capacity to learn combine to make animal communication increasingly rich and complicated.

Not only is the predictive value of the signal enriched by learning, but the characteristics of the signal are also enriched. Nowhere is this more obvious in the animal kingdom than in the song of birds. This is an immense and fascinating field of research and the founder of my own laboratory, Bill Thorpe, can take most of the credit for also founding the field. It is clear that male birds with more complicated repertoires can intimidate rivals more successfully and/or attract mates more readily. The blackbird acquires extremely complicated songs, incorporating noises from what it hears in the environment, such as the songs and cries of other animals. Recently birds have even incorporated the beeps of pelican crossings and the refined noise of the new-style telephone. Many a person has been seen hastening into their house on a fine summer's day by this bit of imitation. The ability to learn such things provides an important step in enriching the signals of animals. It also provides a link to the last topic I want to cover.

LINKS BETWEEN ANIMALS AND HUMANS

Although my brief has been to examine animal communication, I am bound to be asked whether the Darwinian approach can add anything to understanding the connections between what animals do and what we do. Noam Chomsky argues in his chapter that there is no evidence to support the view that human language is designed for communication (in the sense that it is the product of Darwinian evolution). He suggests that the emergence of what he calls the 'language faculty' is a side-effect of the evolution of a complex brain with many specialised functions. Others have stressed, in contrast, the continuities between humans and other animals. To lend plausibility to their view, attempts have been made to teach apes to talk. What can be said about the success of these attempts and the character of the debate?

To be sure, the great apes behave so intelligently and have such a rich social life that it seemed extraordinary to many people that they could not learn to speak. An apparent breakthrough came when the Gardners[17] taught their first chimpanzee, Washoe, the sign language used by deaf and dumb people. Their experiments were criticised because they had not rigorously excluded unconscious cues that might have been provided by the human trainers, as in the case of the famous German horse, Clever Hans. Then other research workers got chimpanzees to perform complicated signalling under much better controlled situations, using plastic shapes which the chimp could pick up, or images back-projected onto a screen which the chimp could press. Some symbols denoted objects such as bananas, apples and tables. Others denoted actions such as eating, walking and swimming. In the kind of yuppy apartment in which they lived, the chimpanzees would signal 'GO SINK' and head off to the kitchen sink.[18]

One experiment conducted by the Rumbaughs took the following form. The chimpanzee scanned a tray of objects and went to a keyboard out of view of the tray and pressed a symbol corresponding to one of the objects. The chimpanzee then returned to the tray of objects, and picked the object corresponding to the symbol. Finally he brought the object to a human who was out of sight round a corner.

In the birds, the brain has evolved so that in some groups it is comparable in size and complexity to that found in primates. Relative to its body size, the parrot has the biggest brain of all. It has complexity of social behaviour to match. Irene Pepperberg[19] has spent a great deal of time training an African grey parrot which she calls Alex (Figure 8). Parrots have an advantage over and above chimpanzees in that they can, of course, imitate human sounds. But, contrary to popular belief, they are not at all analogous to tape-recorders. Alex has been trained to produce vocal answers to questions from a human. For instance, Alex was shown pairs of objects that had nothing in common and asked 'What's the same?' Alex looked at the objects and said 'None'. He was then given two objects that were identical and asked 'What's different?'. Back came the reply 'None'. Now, you might think that is unremarkable and argue that he simply says 'None' to everything. But when Alex is shown two objects that differ in one of three respects: colour, shape or the material of which they are made and asked 'What's different?', Alex says:

Figure 8 The parrot Alex taught by Irene Pepperberg to give verbal answers to spoken questions about two objects (Photo: D. Linden).

'Colour', if they differ only in colour and: 'Shape', if they differ only in shape. If they differ only in what they are made of he says: 'Mahmah', which is his version of 'Matter'. In some ways, the result of her formal experiments are less dramatic than the incidental bits of behaviour shown by Alex. After a long session of testing, Alex became increasingly crotchety and unco-operative and, in one marvellous sequence captured on video-tape, after a tiring day at the office the parrot says: 'I'm going to go away'. He then waddled off and hung his head in a corner. His behaviour is a long way from human language with all its expressive power and complex syntactic properties. Even so, it is a great deal more than would have been credited to a bird even a few years ago. The important conclusion is that *some* of the cognitive capacities needed for language evolved long before humans. It was a necessary condition for the big leap in the ability to communicate that took place with humans, but it was not sufficient.

It is obvious that humans have far bigger brains than even our closest relatives and the fossil record suggests that the rate of evolution has been spectacularly fast, the brain size more than doubling in less than two million years (see e.g. Foley, *Another Unique Species*). This is what one would expect in a process that feeds back positively on itself, further promoting the conditions that got it going in the first place. A fashionable explanation is that this rapid upward spiral is the product of Machiavellian intelligence by which individuals gain great advantage from outwitting others.[20] Alternatively, or in addition, the upward spiral is the product of some of the surprising consequences of co-operation. I have already given examples of different characteristics evolving for different reasons. When these combine they open up a new range of possibilities on which Darwinian evolution can act. This thought may be helpful in thinking about the evolution of human language.

Capacity to learn about associations in the world is clearly of great value in a variable environment and appeared at an early stage in animal evolution. Social life can lead to individuals genuinely working together and to signals that indicate the state of the individual. These signals probably serve to maintain the coherence of the group to the benefit of the individual. They can also provide valuable information about the rest of the environment. In the most complicated animals, social life can lead to transmission of skills by copying. The next step is to enrich the behavioural repertoire used in communication by imitating the signals of others and to associate the performance of signals with the context in which they are given. It is not a big step from here to the symbolic use of signals. In general, if one individual is able to make its future actions plainer than another and if it derives benefit from doing so (which does not seem implausible in a social context), then we should expect evolutionary change in the effectiveness of such signals. All of the capacities could have evolved independently, but then obtained further impetus for change from the gradual emergence of language. All would lead to more computing capacity and larger brains.

My response to Chomsky, then, is that he is right in part to treat the evolution of human language as an emergent property of other characteristics. The question is whether he is wrong to argue that the capacity for language does not carry any benefit to the individual and, therefore, was not subject to further evolution. Once the ability to create symbolic signals and rearrange them was in place, individuals that accomplished this more effectively might well have been at an advantage over those who did it less effectively in terms of all the benefits that come from co-operation. We may eventually be able to check whether the verbal arguments for the Darwinian evolution of language can be simulated on a computer. If and when this is achieved, the proposal would at least seem plausible. However, it will always be difficult to obtain direct evidence for such a view since the less successful members of the lineage left no descendants for us to study.

CONCLUSION

A number of themes have run through this chapter. I began by pointing out that the idea of evolution has been used in two different ways. One is that, because we have evolved from other animals, it makes sense to project downwards from ourselves. The other approach is to look for the advantage to the individual of behaving in a particular way. This second approach relies heavily on Charles Darwin's theory of evolution. However, I argued that we should not suppose that the essentially competitive process he proposed implies a competitive outcome. On the contrary, a great deal of communication is to do with signals that carry real information – even in conflicts – and with co-operation in which all the participants benefit by working with each other. Another popular misconception of Darwinian evolution is that its products must be genetically determined because their inheritance depends on genes. In fact we have seen many examples of animal communication that involve learning. I argued that individuals can gain great advantage from their ability to refine and enrich their methods of communication by this means. Finally, I made the point that relates to the possible links between animal and human communication. As capacities evolved, they may have combined with others that have evolved for quite different reasons to create new characteristics that then acquired an evolutionary life of their own. If this happened, the style of projecting our intentions and emotions into other animals, which we all find so easy and which seems to flow from a belief in evolution, will often lead us astray. We assume continuities of communication between other animals and ourselves where there may be none. Solomon was probably wise not to have tried to speak with animals.

REFERENCES

1 R. Dawkins, *The Selfish Gene*, Oxford University Press, 1976.
2 R. Dawkins and J. R. Krebs, 'Arms race between and within species', *Proceedings of the Royal Society of London*, B 205 (1979), 489–511.
3 J. Maynard Smith and S. E. Riechert, 'A conflicting-tendency model of spider agnostic behaviour: hybrid-pure population line comparisons', *Animal Behaviour*, 32 (1984), 564–78.
4 *Ibid.*
5 P. G. Caryl, 'Communication by agnostic displays: what can games theory contribute to ethology?', *Behaviour*, 68 (1979), 135–44.
6 *Ibid.*
7 R. A. Hinde, 'Animal signals: ethological and games-theory approaches are not incompatible', *Animal Behaviour*, 29 (1981), 535–42.

8 A. B. Bond, 'Toward a resolution of the paradox of aggressive displays: I Optimal deceit in the communication of fighting ability', *Ethology*, 81 (1989), 29–46.
9 A. M. Spence, *Market Signaling: Informational transfer in hiring and related screening processes*, Cambridge, Mass.: Harvard University Press, 1974.
10 J. Farrell and R. Gibbons, 'Cheap talk can matter in bargaining', *Journal of Economic Theory*, 48 (1989), 221–37.
11 T. H. Clutton-Brock and S. D. Albon, 'The roaring of red deer and the evolution of honest advertisement', *Behaviour*, 69 (1979), 135–44.
12 C. ten Cate and P. Bateson, 'Sexual selection: the evolution of conspicuous characteristics in birds by means of imprinting', *Evolution*, 42 (1988), 1355–8.
13 P. Bateson and D. C. Turner, 'Questions about cats', in *The Domestic Cat*, ed. D. C. Turner and P. Bateson, Cambridge University Press, 1988, pp. 193–201.
14 J. L. Gould, 'The dance language controversy', *Quarterly Review of Biology*, 51 (1976), 211–44.
15 D. L. Cheney and R. M. Seyfarth, 'Social and non-social knowledge in vervet monkeys', *Philosophical Transactions of the Royal Society of London*, B 308 (1985), 187–201.
16 W. Vieth, E. Curio and U. Ernst, 'The adaptive significance of avian mobbing. III. Cultural transmission of enemy recognition in blackbirds: cross-species tutoring and properties of learning', *Animal Behaviour*, 28 (1980), 1217–29.
17 B. T. Gardner and R. A. Gardner, 'Signs of intelligence in cross-fostered chimpanzees', *Philosophical Transactions of the Royal Society of London*, B 308 (1985), 159–76.
18 E. S. Savage-Rumbaugh, R. A. Sevcik, D. M. Rumbaugh and E. Rubert, 'The capacity of animals to acquire language: do species differences have anything to say to us?', *Philosophical Transactions of the Royal Society of London*, B 308 (1985), 177–85.
19 I. M. Pepperberg, 'Comprehension of "absence" by an African Grey Parrot: Learning with respect to questions of same/different', *Journal of the Experimental Analysis of Behavior*, 50 (1988), 553–64.
20 A. Whiten and R. Byrne, eds. *Machiavellian Intelligence: Social Expertise and the Evolution of Intellect in Monkeys, Apes and Humans*, Oxford University Press, 1988.

ACKNOWLEDGEMENTS

I am grateful to Robert Hinde, John Lyons and Hamid Sabourian for their comments on the draft of this paper.

FURTHER READING

Bateson, P. (ed.) *Mate Choice*, Cambridge University Press, 1983.
Dawkins, R. 'Communication' in D. McFarland (ed.), *The Oxford Companion to Animal Behaviour*, Oxford University Press, 1981, pp. 78–91.
Foley, R. *Another Unique Species*, Harlow, Essex: Longman, 1987.
Gould, J. L. 'Buzzing Bees' in P. J. B. Slater (ed.), *The Collins Encyclopaedia of Animal Behaviour*, London: Collins, 1986, pp. 70–1.
Halliday, T. R. and Slater, P. J. B. (eds.) *Animal Behaviour*, vol. 2: *Communication*, Oxford: Blackwell, 1983.
Slater, P. J. B. 'Communication' in P. J. B. Slater (ed.), *The Collins Encyclopaedia of Animal Behaviour*, London: Collins, 1986, pp. 62–9.
Thorpe, W. H. *Bird-song*, Cambridge University Press, 1961.
von Frisch, K. *The Dance Language and Orientation of Bees*, Cambridge, Mass: Harvard University Press, 1967.

3

Language and mind

NOAM CHOMSKY

Questions of language and mind have been addressed for millennia, and with particular intensity of focus in the past several decades. A persistent theme is that understanding of language, and human thought and action more generally, has been inhibited by a good deal of intellectual folklore. Many would hold that this remains the case, with varying judgements on where the hazards and pitfalls lie. There is little agreement on fundamental questions. Lively controversy over technical issues is the norm among researchers who share a general perspective. In offering an account of current understanding, I will be giving a personal view, adopting conclusions for which argument has been presented elsewhere without meaning to imply that the issues are settled.

The work that concerns me here has proceeded within the framework of what some call 'the cognitive revolution' of the 1950s. At the time, there were grand hopes and expectations. In the early twentieth century, fundamental physics had been radically modified, leading to hopes for far-reaching unity of science. Quantum theory explained 'most of physics and all of chemistry' so that 'physics and chemistry have been fused into complete oneness...' as the physicists Paul Dirac and Werner Heisenberg observed. The discoveries of early genetics were accommodated within known biochemistry, eliminating the last vestige of vitalism from scientific biology and offering the hope that the evolution and growth of living organisms might fall within the unified natural sciences as well. The next scientific frontier was naturally assumed to be the human mind and its manifestations in thought and action, judgement and evaluation, creation and understanding. The emerging cognitive sciences were to carry us across this frontier, assimilating the study of human thought and action to the core natural sciences. Speaking personally, I did not entirely share these expectations – hopes or fears, depending on one's point of view – and remain sceptical today.

The 'cognitive revolution' has been variously understood by those who participated in it. As it looked to me, the 'revolution' offered a shift of perspective with regard to the problems of language and mind: from behaviour and the products of behaviour (for example, utterances, discourses, and texts), to the inner mechanisms of mind that underlie behaviour and determine its form and character, and how it is interpreted and understood. In the case of language, an alternative was counterposed to the prevailing conceptions framed in terms of habits, abilities, dispositions, skills, patterns and structures. The alternative was a computational-representational theory of mind: the mind, using its internal mechanisms, forms and manipulates symbolic representations and uses them in executing actions and interpreting experience. Communication, from this point of view, is no specific function of natural language; humans have other modes of communication, and there seems no reason to single out communication among the many uses to which language is put.

With this shift of perspective, behaviour and its products are no longer taken to be the object of investigation. Rather, these are just data, of interest insofar as they provide evidence, alongside of other kinds, for the study of what really concerns us: the inner mechanisms of mind. We thus shift to the standpoint of the natural sciences, seeking evidence that will provide insight into hidden realities, abandoning a hopeless search for order in the world of direct experience. We also put aside the picture of language as a system of habits or dispositions to respond, ideas that had been barren of consequences and remain so. Recognising the sharp conceptual distinction between knowledge and ability, between cognitive systems and their use, we can proceed to study structures of knowledge, understanding, evaluation and the like, and to determine their principles, their properties and interrelations, how they develop in the mind, and the ways in which they enter into interpretation and action. In a few domains at least, the effects were soon apparent in novel insights and theoretical ideas, challenging and hitherto unrecognised problems, and a flood of new empirical materials that could be at least partially understood.

It is of some interest that although the conceptions of language in terms of habit, dispositions to respond, and the like, were widely accepted, and indeed still are, there has been little effort to develop and apply them; that is, to suggest how the phenomena of language might be explained or even superficially described in these terms: for example, the relations of form and meaning that have been investigated, often with much success, in other terms. No less surprising is that these ideas are regularly put forth as the austere approach of the no-nonsense scientific mind. This is a serious misunderstanding: it would be more accurate to describe them as relics of a pre-scientific era and to account for the empirical inadequacies and lack of progress in these terms. In the natural sciences, one finds no comparable prior conditions on legitimate theory construction. I think there are interesting questions here concerning the general intellectual culture, but that is another topic.

The avowed mentalism of the cognitive revolution should be understood as a step towards integrating the study of language and other aspects of psychology within the natural sciences. We might think of nineteenth century chemistry as the study of the properties of unknown physical mechanisms, expressing its principles and descriptions in terms of such abstract notions as chemical elements, valence, the structure of organic molecules, the Periodic Table, and so on. This abstract study set the stage for the subsequent inquiry into 'more fundamental' entities that exhibit the properties formulated at the abstract level of inquiry. The same may be said of early genetics. Correspondingly, the study of computational-representational theories of mind and their role in action and understanding should serve as a guide for the brain sciences of the future, providing them with an analysis of the conditions that the mechanisms sought must satisfy. Thus when I use such terms as 'mind' or 'mental', I mean nothing that should be at all controversial: the reference is to the study of conditions on as yet unknown physical mechanisms, a typical feature of the natural sciences.

Suppose that in the course of time, as we hope and anticipate, neural mechanisms are discovered that exhibit the properties and satisfy the conditions formulated in terms of such entities as the rules and representations of language. We will not conclude that these entities do not exist, any more than the unification of chemistry, parts of biology and physics show that there are no chemical elements and ions, genes and alleles, continents and galaxies, plants and animals, or persons understood in the highly abstract terms in which we conceive them. Some of these theoretical constructions may be shown to be misguided, going the way of phlogiston and vital forces; others may be sharpened and modified as understanding moves forward.

But generally, we expect, progress in establishing the links among various levels of inquiry will provide better understanding of entities of which we have only a partial grasp when limited to only certain levels of theory construction and explanation. It would also come as no great surprise if the 'fundamental sciences' had to be modified to accommodate the properties unearthed in the study of mind, as has often happened in the past. And it remains, as always, an open empirical question whether such modifications, if necessary, lie within the scope of human intelligence, which is after all a particular biological structure and not a universal system available for all contingencies.

Talk about mind, then, is talk about the physical world, but not necessarily the physical world as it is now assumed to be. We can say that mentalistic formulations are about the physical world, but only if we understand the physical world to be whatever there is, in which case we are merely saying that we propose these formulations as possibly true. There is no fixed notion of the physical against which we can compare formulations in other terms; there is only our best guess as to the facts, formulated in whatever terms we can put forth for empirical evaluation. There is no 'mind' as distinct from 'body', because we have no clear and fixed conception of body, and have had none since the collapse of the intuitive mechanics of the seventeenth century. There should, then, be nothing controversial in the mentalistic formulations of computational-representational theories of mind, apart from the question of their truth.

The 'cognitive revolution' of the 1950s was not as novel as many assumed it to be. In significant respects, it recapitulated leading ideas of what we might call 'the first cognitive revolution' of the seventeenth century. Descartes and his followers also developed a computational-representational theory of mind, extended in interesting ways into the nineteenth century. This work raised and sometimes partially answered questions that were revived during the 'second cognitive revolution'. This was particularly the case in the domains of vision and language, just those where progress has been most rapid since the revival of these ideas.

We can extricate three fundamental problems from the classical tradition. The first we might call 'Humboldt's problem'. Wilhelm von Humboldt recognised early in the nineteenth century that language involves 'infinite use of finite means'. He regarded language not as a collection of constructed objects – utterances or speech acts – but rather as a process of forming structured expressions. With a bit of interpretative licence, we might understand him to be saying that a language is a generative procedure that enables articulated, structured expressions of thought to be freely produced and understood. The interpretative licence lies in the fact that Humboldt did not, and at the time could not, distinguish clearly between a generative procedure that determines the properties of linguistic expressions, and the mental process that brings thought to expression in linguistic performance. With later advances in the formal sciences, this crucial distinction becomes clear, and we can confront empirical problems that could only be intuitively sensed prior to this conceptual clarification.

I will understand 'Humboldt's problem' to be that of determining the finite means that are put to unbounded use. In the classic view of Humboldt and Sapir, to *have* a language or to *know* a language is to have mastered a certain way of speaking and understanding, which means, at least, to have acquired the finite means that are freely put to use. To the best of our current understanding, this means that to know a particular language is to have encoded in the mind/brain a certain generative procedure, an algorithm of the kind that one might program for a computer, which assigns a specific interpretation to every possible linguistic expression. Humboldt's problem, then, is to characterise these procedures in particular cases, thus giving an account of knowledge of particular languages.

In technical terminology, the encoded computational procedure *strongly generates* an infinite set of *structural descriptions*. Each structural description specifies the properties of a particular linguistic expression: its sound, its meaningful and grammatical elements, their arrangement into phrases, structural connections among the parts, and the meaning of the expression insofar as it is determined by the internally represented cognitive system. Consider, for example, the following two expressions:

(i) he thinks that John is intelligent
(ii) his mother thinks that John is intelligent

The generative procedure encoded in my mind/brain assigns to (i) a phonetic representation; an analysis into such units as *he, think, be, present tense* (occurring twice), etc., and a hierarchy of phrases; and so on. The structural description will also indicate that the reference of the pronoun *he* is determined by the discourse situation, not by a structural relation to *John*, taken as its antecedent. In the case of (ii), the structural description would be the same, apart from the analysis of the phrase *his mother* and the absence of the condition on reference of the pronoun. We can use (ii) to express the thought that John's mother thinks that he (John) is intelligent, but we cannot use (i) to say that John thinks he (John) is intelligent.

Let us call the class of structural descriptions determined by the generative procedure the *structure* of the language in question. Then to know a language is (at least) to have mastered a procedure that determines the structure of that language. The mind/brain has reached a state in which this generative procedure is encoded in physical mechanisms that are yet to be discovered.

The least controversial conclusions about language of any significance are the ones just sketched. I know of no sound reason to doubt the validity of this conception, nor of any coherent alternative proposal regarding the ability to speak and understand. Furthermore, the range of explanatory success is not inconsiderable.

Adopting this point of view, what will we take a language to be? We might understand a language to be nothing other than the internalised generative procedure. Elsewhere, I have suggested that we use the term 'I-language' to refer to language construed in this way, where 'I' is to suggest 'internalised' and 'intensional'. The I-language is 'internalised' in that it is encoded in the mind/brain, so that our study of it is a normal part of the natural science; it is 'intensional' in that it is a specific method for constructing structural descriptions (technically: a function taken in intension). The I-language is thus analogous to a specific procedure for determining integral square roots, a particular program that one might write for a computer to carry out this task. The structure generated is analogous to the infinite list of pairs (of numbers and their square roots) – (1, 1) (4, 2) (9, 3) ... – taken in extension, apart from any specific procedure for calculating them.

One may ask whether there is any other notion of language coherent enough, or useful enough, to be seriously considered for the study of language and mind. Whatever the conclusion, the legitimacy of the concept of I-language, and the associated notions of strong generation and structure, does not appear to be in doubt.

In the technical literature, the term 'grammar' has been used to refer either to I-language or to the linguist's theory of the I-language. To avoid misunderstanding, we may restrict the term 'grammar' to the linguist's theory, in the spirit of traditional usage. A grammar, then, is a theory of an I-language, the latter a generative procedure encoded in the language faculty. Humboldt's problem is the problem of constructing grammars.

There is much debate over whether it is proper to say of Jones, who knows English, that Jones knows the grammar of English. If by the phrase 'grammar of English' we mean the linguist's theory of Jones's I-language, then the answer is surely 'No'. If by the phrase 'grammar of English' we mean the I-language that Jones has, then the answer to the question is largely a matter of decision. Do we want to use the term 'knowledge of language' to refer to the property of having an I-language? Is this an appropriate rational reconstruction of the concept of knowledge? Do we want to say further that if some principle of language – say, the principle of *binding theory* that determines what the reference of pronouns depends on – is a component of the I-language , even an innately determined component, then Jones knows this principle, so that we explain the uncontroversial fact that Jones knows that such-and-such in terms of his knowledge of the principles from which the fact derives?

Little of substance is at stake. Clearly, the notion of knowledge that will be constructed in this way is not identical to that of normal discourse, though it is, I think, closer to that notion than familiar theoretical constructions in terms of ability and capacity, for reasons I have discussed elsewhere. Equally clearly, the term 'knowledge' of ordinary usage, with its open texture and variety of uses, is as unlikely to survive the transition to explanatory theory unchanged as any other term of common sense discourse. The important question is whether the concept developed in this way is one, or *the* one, that we want if we hope to gain understanding and insight into phenomena that are called instances of knowledge in informal discourse: knowing that such-and-such an expression means so-and-so, or consists of certain words and phrases but not others, for example. To advance the discussion, what we need is sharpened conceptions and theories, and evidence of their success in providing explanation and understanding.

Let us consider some other proposals regarding the nature of language. In the formal sciences, a language is taken to be a class of well-formed expressions; we may select one or another generative procedure to specify them, the choice being a matter of convenience rather than principle. We say that the selected procedure *weakly generates* the set of well-formed expressions. We might refer to this set as an 'E-language', where 'E' is to suggest 'externalised' and 'extensional'; the E-language is 'external' to any mind/brain, a certain infinite set regarded in extension, like the class of pairs (x, y), where y is the integral square root of x. In studying formal arithmetic, for example, given a notation we select a set of expressions as the E-language, the set of well-formed expressions. In one standard notation, the expressions '(2 + 2) =4' and '(2 + 2) = 11' are well-formed, but '2) + 2(4 =' is not.

We then select one or another generative procedure to provide a specification of the E-language. It is the E-language that interests us; the specific procedure (the I-language) that generates it does not.

However appropriate for the study of formal systems, the notion of E-language apparently lacks redeeming value for empirical inquiry. There is wide appeal to such a notion in the literature of linguistics, philosophy, and psychology, but I know of only two attempts to try to clarify what the class of well-formed expressions might be, neither successful. One is a theory-internal construction of my own thirty-five years ago, soon abandoned. The other is a behavioural criterion suggested rather vaguely by the Harvard philosopher W. V. Quine, which seems of no particular promise. Though the notion of E-language is unproblematic (by stipulation) in formal language theory, its empirical status is quite dubious.

Until some sense is given to the notion of E-language, we have no way of understanding what people mean when they talk about 'extensionally equivalent grammars' (as in efforts to give some content to Quine's notion of 'indeterminacy') or when they tell us that languages have, or lack, some formal property; say, that they are, or are not, recursive (that is, there is, or is not, a precise way to determine whether an expression is well-formed in the language). For this reason alone we may dismiss a large literature on the alleged implications of the weak generative capacity of grammars for language processing or language learning. Evidently, we can make no sense of such assertions if we do not know what an E-language is supposed to be. For the present, there is little reason to suppose that E-languages are generated at all, and even if such a notion can be developed and introduced into the study of language, it is unclear why we should care what formal properties these objects would turn out to have.

But we do have one useful notion that is extensional and externalised: the notion of structure. Study of the formal properties of structures might have some consequences for our understanding of language processing and acquisition. One suggestive example is the strong equivalence of non-deterministic pushdown storage automata and context-free grammars. The formal study of various classes of E-language has had some indirect empirical relevance, by clarifying properties of language that cannot be captured by generative procedures of certain types. There are also suggestive results in formal learning theory and complexity theory. But there have also been serious misunderstandings about all of these matters, with most unfortunate effects on empirical research.

Notice that *specimens* of E-language do exist, that is, particular utterances, or speech acts, or other bits of behaviour. The question at issue is whether some infinite set of these objects has a special status in human psychology. To illustrate some of the problems that arise, consider the English sentence:

(1) they think [that] Mary fixed the car with a wrench

We can inquire into what Mary fixed and how she fixed it, forming the interrogative expressions:

(2) what do they think [that] Mary fixed with a wrench? (Answer: *the car*)
(3) how do they think [that] Mary fixed the car? (Answer: *with a wrench*)

These expressions are called 'well-formed' and 'grammatical'.

Consider now the sentence:

(4) they wonder why Mary fixed the car with a wrench

Forming interrogatives as before, we derive the expressions:

(5) what do they wonder why Mary fixed with a wrench? (Answer: *the car*)
(6) how do they wonder why Mary fixed the car? (Answer: *with a wrench*)

The first of these is somehow deviant, and the second, virtual gibberish; we can interpret (6) only as a query about how they wonder. This expression is called ungrammatical in the interpretation in which it is concerned with how the car was fixed.

In short, extraction of *what* or *how* from the *that*-phrase yields a grammatical expression and extraction of *how* from a *why*-phrase yields an ungrammatical sentence; the former two cases are taken to be 'in the E-language' and the last is not. But what about extraction of *what* from the *why*-phrase (example (5)? The resulting expression differs in status both from the grammatical expressions (2) and (3) and from the ungrammatical expression (6) (in the intended interpretation). Is (5) grammatical, ungrammatical, something else? And what about the expression (6) interpreted as a query about how you wonder? It too differs in status from the grammatical expressions and the ungrammatical ones, and also from the deviant example just given, because it is not clear what is meant by a way of wondering. So now we have four categories with regard to well-formedness, and we can easily proliferate such categories. Where, then, is the boundary of well-formedness?

From the very origins of work in generative grammar it has been recognised that these are the wrong questions. Each of these expressions has a structural description assigned to it by the I-language, and is interpreted in terms of the structural description assigned; the same is true of expressions of Swahili, which are surely heard and interpreted differently by monolingual speakers of English and Japanese, by virtue of the I-languages they have internalised: in fact, we could learn a good deal about my I-language by studying how I interpret expressions of Swahili. It is not clear that there is any principled reason to define an E-language with the bounds of 'well-formedness' set in one or another place. There is, in short, little reason to believe that such a concept has any empirical status.

Such questions arise over a very wide range. Is the expression 'misery loves company' grammatical, or is it not? Plainly it is somehow deviant; otherwise it would lose its force as a rhetorical device. But it is immediately understandable, readily parsed, appropriately used, and assigned a determinate structural description by the I-languages of speakers of English.

In these respects, natural languages are quite unlike formal arithmetic. In the latter, there is a sharp and precise boundary between well- and ill-formed: misplace a parenthesis, and the expression has no interpretation at all. Natural languages do not work that way.

I might add that data such as those illustrated have long been critical in the study of language. The difference in status between extraction of *what or how* from a *why*-phrase, both deviant but in quite different ways, turns out to be much more interesting than the difference between the unproblematic cases of well-formedness. This becomes still clearer if we consider other possible interrogative constructions based on the same forms, for example, (7) and (8):

(7) who do they think [that] fixed the car with a wrench?
(8) who do they wonder why fixed the car with a wrench?

Sentence (7) is normal in some varieties of English, deviant in others. In contrast, (8) is sharply deviant in all varieties of English and similar languages.

Questions such as these lead us directly to fundamental features of the computational systems that determine form and meaning. The data are therefore of real significance. They bear on some of the most lively and important issues of current research. The interest of data is a theory-relative matter, and in the present state of understanding, the various forms of deviance happen to be of critical importance for determining the principles of language. Such research could not even be pursued if we were saddled with the dichotomy between well-formed and ill-formed, borrowed from the formal sciences. It seems that we can safely – and wisely – dispense with the notion of E-language.

A common misunderstanding is that the goal of the theory of generative grammar is to construct grammars (of I-language) that distinguish grammatical from ungrammatical expressions, such expressions constituting the data for the field. So viewed, the theory would appear to be a refinement of structural and behavioural linguistics, which sought to find the elements of language samples and their patterns and arrangements. But this is a complete misconception. From its origins, generative grammar rejected any absolute notion of well-formedness, regarded patterns and arrangements as basically epiphenomena or descriptive conveniences, and sought to find the inner mechanisms that determine form and meaning, a completely different task.

Another familiar concept of language is framed in terms of communities, norms, conventions, social practices, and the like. As the idea is formulated by the Oxford philosopher Michael Dummett, in its 'fundamental sense', language is a community property that exists 'independently of any particular speakers', a social practice of which each speaker has only a 'partial, and partially erroneous, grasp'. This conception dominates philosophical analysis and is widely adopted elsewhere too, though in the empirical disciplines that actually study language the notion is scarcely to be found, except possibly as a high-level abstraction of some sort. Can any sense by made of this notion?

The question is not whether the concept of language as a community property satisfying certain norms and conventions is useful for ordinary discourse. Doubtless it is, and we need not dispense with it any more than we abandon talk of the rising of the sun. Adopting this informal notion, we can say, of the child Mary, that she has only partial knowledge of our language; she doesn't yet know the meaning of the words 'tendentious' and 'sententious', as some of us may. Similarly, if Jones uses the word 'disinterested' to mean 'uninterested', we can say that he is making a mistake; the 'common language' requires a different usage. Varying interpretations of all of this can be given in terms of authority structures, individual aspirations, political units, and the like. There is, however, no way to draw boundaries corresponding to such languages in any non-arbitrary way, no basis for legitimate or useful idealisation, and little reason to believe that such notions can be clarified in a way that will contribute to the study of language and mind.

Notice that we may construct various *expansions* of Mary's I-language, adding on something that is not now a part of it. Thus one expansion would include a definition of *sententious* as 'given to excessive moralising', and another would define the word as 'excessively sentimental'. Either of these expansions is a possible human language. There are many expansions of Mary's I-language that would constitute a possible human language, and many others that would not, as a matter of biological fact. We might say that Mary has only partial knowledge of all the legitimate expansions, but there is no particular subset of them that has any privileged existence from the point of view of Mary's psychology and biology, though some surely do in the world of historical and social accident. As for Jones, who uses 'disinterested' to mean 'uninterested', he might be in error if his own I-language happens to yield a different consequence; many factors interact in behaviour besides the internalised generative procedure, so our behaviour might not correspond to our knowledge. But apart from that, for the purpose of the cognitive sciences at least, there is no useful sense in which Jones is violating the norms of some 'common language', English, just as I am not misusing this common language when I pronounce words in my urban Philadelphia accent.

These remarks should be truisms. Furthermore, it is by no means clear how to salvage any significant notion of 'misuse of language', 'partial knowledge', 'common language' or 'conventions and norms'. There is, again, a substantial literature appealing to these notions as if they have some moderately clear sense. They do not, and it is not clear that there would be much point in trying to sharpen them in one or another way. I think the implications of these facts are more far-reaching than is often appreciated, but will not pursue the topic here.

What is reasonably clear is that the mind achieves a state in which it incorporates an I-language. Under these circumstances, we may say either that the person 'knows' or 'has' this I-language. Two people can communicate if their I-languages are close enough; it is a matter of more or less, not yes or no, and there are many dimensions. There is no such object as a 'common meaning' that people grasp or partially grasp, any more than there is a 'common pronunciation' or a 'common shape' that they partially have. I-languages are more or less similar, as shapes are more or less similar (apart from the digital character of the array of I-languages, a matter not relevant here). The notion of common language and those relating to it (misuse, convention, etc.) are vague and scarcely coherent, probably beyond repair, though fine for the purposes of ordinary life. It is the task of people who believe otherwise to give some account of these notions, and to show the point of doing so. That task has yet to be seriously undertaken.

Suppose that Jones has a particular I-language; that is, his mind/brain has achieved a particular state. Is there some point in constructing an abstract object external to Jones that is an image of his I-language, fully determined by it, which we may call Jones's language in some other sense? And further, in constructing a cognitive relation that holds between Jones and this external abstract object, a relation that we might call 'knowledge'? It is hard to see the purpose, either for science, or for clarifying our discourse and thinking. If we construct these new concepts, will there be something that we will then be able to describe perspicuously or explain that we could not readily account for in terms of I-language? The burden of proof is on those who think there is. If it can be met, nothing changes in what I will say within the bounds permitted by the concepts of I-language and having (or knowing) an I-language.

So far I have discussed just one fundamental problem: what I called 'Humboldt's problem'. It deserves more careful thought than it often receives. When we consider it closely, we discover, I think, that we are left with the relatively clear and uncontroversial notion of I-language, a real object, and having an I-language, a real property of real people, but possibly no other notion of language.

The second of the three classical problems is to explain how knowledge and understanding are acquired, what is sometimes called 'Plato's problem'. In our special case, the problem is to determine how the mind/brain reaches the state in which it incorporates a particular I-language. One of the achievements of the second cognitive revolution has been to provide some rather far-reaching answers to this problem, along lines that might not have surprised Descartes, possibly even Plato. It seems that the initial state of the mind/brain, a genetically-determined property, permits a restricted class of I-languages, and that limited data suffice for selecting among them. The transition from the initial state of the language faculty to a state incorporating an I-language has little resemblance to what is ordinarily called learning, a concept that may well prove dispensable. The theory of the initial state of the language faculty is sometimes called 'universal grammar'. I will return to some recent ideas about it and some of their consequences.

In the phenomenal world of social and historical accident, no person ever comes to have an I-language; rather, some hopeless and uninteresting amalgam, for me, a mixture of influences from Philadelphia, Boston, the Ukraine, a variety of authority figures and texts, and who knows what else.

Universal grammar, understood now as part of human biology, specifies the class of systems that a person would acquire in an ideal homogeneous speech community. We may think of it as a *language acquisition device*, that is, a procedure that operates on experience acquired in an ideal community and constructs from it, in a determinate way, a state of the language faculty. Universal grammar is, then, a function that maps experience in an ideal community to an I-language, much as the function *square root* maps 16 into 4 and 49 into 7; again, we are concerned with the function taken in intension, that is, with a specific characterisation of the procedure, encoded in the initial state of the mind/brain.

It can scarcely be doubted that this function exists as a real property of the human mind/brain and that it is put to use in actual language acquisition. To doubt that this is so would be to hold either that language can be acquired only under conditions of varied and conflicting evidence, or that the function exists but is never used in acquiring a language; these ideas are not very attractive. Note that the situation contrasts sharply with efforts to make some sense of the notion 'common language'. Here, there is no basis for legitimate idealisation and no real world entity that can be identified, except in the limiting case of a homogenous community, in which case the notion 'common language' reduces to the study of I-language within the framework of individual psychology.

Actually, still further idealisations are required if we hope to derive a realistic account of the mind/brain, but I will leave the matter here.

The last of the three classical problems might be called 'Descartes' problem', the problem of how knowledge of language is put to use. Here we may distinguish the production problem from the perception problem, and we may put the first aside; there is little to say about its essential properties. The perception problem has received a good deal of study. In its most general terms, the problem is to construct a model of a person who assigns an interpretation to a presented expression in a particular situation. Following Donald Davidson, let us refer to this model as an 'interpreter'. It is at once clear that virtually nothing can be said about this problem; in the situations of actual life, virtually any information and strategy might be relevant to determining what a presented utterance means, or what its speaker may have had in mind. If we want to proceed, we will have to put forth some ideas about the components of the mind/brain that carry out the process of interpretation, abstracting from the complexities of the world of experience, as in any attempt to determine the real elements of the world and their properties.

One standard assumption is that there is a component of the mind/brain that considers only formal properties of the expression itself and that accesses the I-language to assign some kind of structural description to a presented expression; call it a 'parser'. While there can be little doubt of the existence of the I-language as a real psychological object, the concept of a parser is more problematic. Notice that the status of the I-language as compared to the parser is the opposite of what is commonly assumed in the literature, which generally takes I-language (or 'grammar', in the more familiar usage) to be more problematic than the notion of a parser. But there is rather substantial evidence for the existence of I-language – whereas evidence for parser is slimmer and less conclusive. It could turn out that there is no such object, but only an 'interpreter' in Davidson's sense, hence nothing much to say about the perception problem. Probably too pessimistic a guess, though it is only recently that clear evidence to the contrary has been accumulating.

To summarise, we have three classical problems, now reformulated in the terms of a mentalistic approach of the kind we would expect to find in the natural sciences. One problem is to determine the initial state of the mind/brain, specifically its language faculty. The second is to determine the properties of various steady states attained, the particular I-languages that people may have or know. The third is to determine the properties of a parser, if such exists. Other problems, still remote, are to explore the production of language, a special case of the problem of choice of action generally, and to find the physical mechanisms that have the properties that come to light in these abstract inquiries. Beyond this, we move to other domains with other problems and concerns.

Tentatively assuming all of this, let us now turn to some broader issues. It is widely assumed that language must be designed for ease of use and acquisition. Thus it is commonly held (e.g. by the linguist Gerald Gazdar), that 'sentences of a natural language can be parsed' and that parsing is 'easy and quick', a fundamental fact about natural language and a condition on language design. It follows that a theory of language must deal with and explain this fact. Another common belief is that natural languages are readily learnable under normal conditions; that there is a fixed learning procedure that will fix upon any of these languages in a finite time, given appropriate data. Again, it is argued that a theory of language must be responsive to this empirical condition.

Such conditions of parsability and learnability may seem obvious, but they are not. Consider first the parsability condition. One problem is that it is formulated in terms of the notion of E-language, to which no sense has been given. But let us put that aside and grant that we have some criterion of well-formedness, so we now understand the phrase 'sentences of the language'. The claim is that these can be easily and quickly parsed. That, however, is simply not true, by any criterion of well-formedness; hence these considerations cannot pose a condition on language design. There are well-known categories of expressions that are perfectly well-formed and meaningful by whatever criterion one likes, but that cannot be readily parsed, or that are normally parsed incorrectly. (Some of these reflect core features of language design, such as iterated embedding.) It is also a familiar fact that sentences may be assigned the wrong interpretation, or no intelligible interpretation, because the interpreter makes the wrong guesses; what are called 'garden path sentences', such as 'the horse raced past the barn fell', typically parsed as consisting of the sentence 'the horse raced past the barn' with the unattached word 'fell' added on, yielding a highly deviant expression, though in fact it has a well-formed interpretation analogous to 'the car driven past the barn fell'. There are well-known classes of expressions, even fairly short ones, that are well-formed, unambiguous, and meaningful by any reasonable criterion but that can be interpreted only with care and effort. In contrast, expressions that are not well-formed by any criterion are often easily, quickly, and unambiguously parsed, and quite properly used in certain circumstances.

We also sometimes find that the design of language makes it difficult to say what we mean; the structure of language does not allow direct expression of our thoughts. Thus, suppose I know that John wonders why Mary fixed the car somehow, and I want to ask how. I cannot proceed in the straightforward way, following the analogy that works elsewhere and forming the interrogative expression: 'how does John wonder why Mary fixed the car?'. Some circumlocution is required, for reasons deeply embedded in language design. There are many similar cases.

There is no general relation between parsability and well-formedness, by any criterion. Standard assumptions about these questions are simply untenable.

It is true, of course, that the sentences of a natural language that are used in practice are easily parsed, but that is not very surprising. All we can say at this level of generality is that the design of natural language provides many usable expressions, alongside of many unusable ones. The usable ones are available for use. Not a very interesting conclusion, and one that would hold as well for many systems that violate the design properties of natural language. The normal properties of language that render it unusable do not impede communication particularly; the speaker keeps to those aspects of language that are usable, and these are the only ones that the interpreter can process. Nothing in psychology or biology suggests that language design should be different in these respects.

In short, from the fact that languages are successfully used, we can draw only very weak conclusions about their design properties, and conditions on language design that are often taken for granted are not only far from obvious but in fact plainly do not hold.

Similar questions arise about learnability. It is not a logical necessity that language design guarantees learnability. Some natural languages can be acquired, including those that *are* acquired; again, not an interesting conclusion. The language faculty might permit many languages that cannot be acquired and therefore will not be instantiated in practice. In fact, it seems to be the case that the natural languages made available by the language faculty are learnable on the basis of simple data, but that, if true, is an empirical discovery, and a rather surprising one. It is only in the context of recent work within the so-called 'principles-and-parameters' framework that this conclusion seems particularly plausible, I believe.

The theory of evolution gives us little reason to believe that the human language faculty developed in such a way that languages are somehow 'designed for use'. We must not succumb to what the biologists Stephen Gould and Richard Lewontin call the 'Panglossian fallacy', the assumption that each trait of an organism is selected so as to yield near-optimal adaptation to the environment. Many factors beyond survival value enter into biological variation and evolution. 'In some cases at least, the forms of living things, and of the parts of living things, can be explained by physical considerations', D'Arcy Thompson observed in a classic work, and such ideas might reach quite far towards an explanation of the properties of complex systems of nature. What we would suspect on the basis of human biological success is that parts of language are usable; those parts are used, others not. And indeed, that is exactly what we find.

It might turn out that the mind of some organism permits a class of languages or other cognitive systems *none* of which can be acquired or used; if so, indirect evidence would be needed to show that the faculties exist. Nothing in modern biology excludes even this possibility. What we would expect, in such a case, is that the system in question would be 'explained by physical considerations', as derivative from something else. We know of cognitive systems that are available but unused through virtually all of human history. Take the number faculty, unused for most of history, up to the present in some societies. It cannot be that the number faculty, apparently a unique human possession, was specifically selected; rather, it must have developed as a by-product of some other system, possibly the language faculty. There could also be faculties that arise in this manner but cannot be accessed. Stranger things have been found in the physical world.

Considerations of this sort bear on the possibility of providing so-called 'functional explanations' for properties of language. If we construct two systems at random, one a generative procedure that strongly generates structural descriptions, the other a parsing system, we are likely to find some respects in which the two are well adapted to one another, others in which they are not. If the generative procedure is incorporated in the parser, which has access to it for performance, then the parser will be able to make use of the information provided by the generative procedure to the extent that the two systems are mutually adapted. It would be a mistake to conclude that the generative procedure was designed for use by the parser just on the basis of the fact that there is a domain of adaptation. One would have to show that this domain goes beyond what might be expected on other grounds, not an easy task. (What is the space of possibilities that we sample, to ask just the most obvious question?) Note further that functional considerations are not given in advance. There are many possible considerations, and they often conflict. Thus, given some formal property, we can often find some functional consideration it might satisfy. Such problems arise whenever functional accounts are offered, and they are not so easily dealt with.

A closer look raises more subtle questions about form–function compatibility. With these in mind, let us consider further ideas about the generative procedures of the language faculty, some reasonably well supported, others more speculative.

A language, construed as an I-language, has two basic components: a lexicon and a computational system. The lexicon consists of substantive elements such as nouns and verbs, and functional elements such as inflections (tense, agreement, etc.). Perhaps surprisingly, there is more than one possible language; a Martian scientist observing humans might regard this as a curious discovery, though the empirical contingencies of language acquisition would lead him to recognise that the variety must be sharply constrained.

Putting aside the question of how structures are related to phonetic form, there is, I think, mounting evidence that there are no rules that are specific to particular languages or to particular constructions such as interrogatives, passives, and so forth. Rather, general and invariant principles of the language faculty interact, sometimes in intricate ways, to determine the form and meaning of expressions. Associated with these principles are certain options of variation (parameters), each with a small number of possible choices (values), perhaps just two. A particular I-language is determined by a selection of values for these parameters and this choice of lexical items. Apart from lexical choice, to acquire a language, a child must set the values of the parameters; and knowing these values, the linguist should be able to predict the full range of interpretation for expressions with the selected vocabulary. From one setting, we should be able to deduce Hungarian, from another, Swahili. This is an awesome task, far from accomplished. But for the first time, the general contours are becoming clear and there are promising steps towards answering a range of questions that are quite new. My own feeling is that the long and rich tradition of study of language may be entering a new and qualitatively different phase, perhaps with significant broader consequences.

One possibility suggested in recent work is that the computational system is fixed and invariant. In this sense, then, there is indeed only one human language; the variations lie in the lexicon. If so, we would expect three kinds of possible variation: in properties of functional elements, of substantive elements, and of the lexicon as a whole.

Languages do vary in the choice and character of functional elements. It has been plausibly argued, for example, that the agreement system is 'weak' in English and 'strong' in Italian and French, with a further difference in strength between French and the other Romance languages. A variety of empirical consequences ensue. Exactly what parameters enter into these variations is not entirely clear. It has been argued that some languages lack functional elements altogether, Japanese being one suggested case.

There is also variety in the global properties of the class of all lexical items. Thus the relation between a lexical item and the phrases semantically related to it is directional: left-to-right in English, so we have '*hit* the boy', '*claim* that it is raining', '*in* the room', '*proud* of John'; and right-to-left in Japanese, so we have the mirror-image throughout. Typically, the directionality property is fixed for the language, holding of lexical elements of all types. Properties of functional elements and of the entire lexicon have large-scale and often intricate effects on the structures generated.

Turning to the third possibility, the class of substantive elements, some subtle questions arise. It is not clear what options there are within this class. It is possible that there is no variation that affects the way in which the computational system operates and no parametric variation of the kind found among functional elements. Of course, there is a choice as to the phonological form assigned to a concept (*tree* in English, *Baum* in German, etc.). And some substantive elements (the causal element, for example) may appear in some language as an independent element, as in English, or an affix, as in Japanese. It might be that there is a fixed and invariant store of lexical concepts from which the person acquiring the language draws elements. The rate of acquisition can be extremely rapid. At peak periods lexical items appear to be acquired on virtually a single exposure. These and other considerations provide compelling reasons to believe that concepts enter the lexicon with a rich and determinate structure, which yields an elaborate array of semantic and other connections among linguistic expressions (including, in particular, so-called analytic connections that determine necessary truths). The functioning of the computational system yields other such connections, and the entire array, reflecting invariant properties of faculties of the mind/brain, provides an *a priori* framework for thought and its expression in natural language. Commonly accepted alternatives in terms of centrality of belief appear to be wholly implausible, a fact that once again has broad consequences.

A reasonable conclusion, then, is that acquisition of a language reduces to selection of substantives from a given store and fixing of values of parameters that apply to functional elements and to properties of the lexicon as a whole. Such a conclusion is far from having been established, but in the areas where there is some clarity about these matters, it seems plausible. If so, then the formal and semantic properties of linguistic expressions and a network of relations among them are substantially determined by the human language faculty itself, in common for all natural languages.

There is reason to believe that the computational system provides three fundamental levels of representation: D-structure (deep structure), PF (phonetic form), and LF (logical form, to be understood with familiar qualifications). Each of these levels stands at an interface of the computational system and some other system of the mind/brain. D-structure directly reflects properties of the lexicon; it expresses these properties in formal structures through the medium of a highly restricted type of phrase structure representation. PF is the interface with the motor-perceptual system. LF is the interface with other cognitive faculties of the mind/brain.

The structural description of an expression consists of a representation at each of these three interface levels. Each element of these representations must be *licensed* by an appropriate relation to the external system at the interface. There can be no 'superfluous symbols' not externally licensed in this fashion. We therefore have a condition of 'economy of representation'. Note that the condition is not true *a priori*. Formal systems invented for practical purposes often violate it, for example, with such devices as vacuous quantification. In standard formal systems, the expression 'for all x, $2 + 2 = 4$' is well-formed, meaning that $2 + 2 = 4$, with the quantifier 'for all x' vacuous. But natural languages, it seems, satisfy the economy condition. The sentence 'who John saw Bill' does not mean that John saw Bill, with the quasi-quantifier 'who' vacuous, and 'every someone left' does not mean 'someone left', with 'every' vacuous. A theory of grammar that provides a technical apparatus to ensure that these consequences hold is disconfirmed, since the consequences follow from general principles, even without the extra (hence unmotivated) apparatus.

The condition of economy of representation is met by definition at D-structure, which is a projection of lexical properties. The condition has always been tacitly assumed as the level of PF. Thus if some phonological representation contains symbols that do not have a language-independent interpretation in motor-perceptual terms, we do not consider it a phonetic representation, but rather a 'higher level' representation that must still be converted to a phonetic form by some I-language computation. The condition at the level of LF has an interesting range of empirical consequences. To mention one case, some recent work suggests that this condition at LF is part of the explanation for the differences of status among the interrogative expressions (2)–(3) and (5)–(8), discussed earlier.

A basic property of language design is that the three representations of the structural description are not directly related to one another; rather, their relations are indirect, mediated by a fourth level: S-structure (surface structure). From this point of view, S-structure is a derivative level. An S-structure representation is something like the solution to a certain set of equations; given the three fundamental representations, each externally licensed, the S-structure representation is selected so as to mediate among them by the mechanisms permitted in the computational system. Presumably, the solution to this set of conditions is unique.

The relation of S-structure to PF and to LF is commonly assumed to be directional, that is, a step-by-step computation from S-structure to PF on the one hand and to LF on the other. The relation of S-structure to the lexicon has been more controversial. If it too is a directional mapping, the D-structure is projected from the lexicon and converted into S-structure by a computational procedure. Note that these are tricky issues. There is, at best, a subtle difference between a relation and a directional mapping. But it seems possible to give some empirical content to this distinction, and I think the evidence currently available tends to support the picture just outlined.

If this is correct, then the relations among the three externally licensed levels are determined by an actual derivation, a step-by-step computational process. The computation converts the lexicon into a pair of representations, one of which is an interface with motor-perceptual systems, the other an interface with other cognitive systems.

These computations have to meet still another overarching condition. Each derivation must be *minimal*, in a certain well-defined sense. Suppose a D-structure representation to be given. From it, we compute a representation at S-structure, then LF. These two representations must be reached by the 'least costly' derivation from D-structure. A most costly derivation, yielding a different choice of S-structure or LF, is illegitimate. To first approximation, cost is determined by counting the number of steps in a computation. But more than this is involved. It seems that invariant principles of language count as 'less costly' than those that reflect particular parametric choices. Intuitively, it is as if the invariant principles are 'wired in' and therefore less costly than language-specific properties. Note that if this is correct, then in the steady state of mature knowledge of language, there must still be a differentiation between invariant principles of the biological endowment and language-specific principles.

There is, again, a wide variety of empirical consequences to these assumptions about language design, and they seem to be confirmed over an interesting range.

To summarise, a valid structural description consists of three fundamental levels of representation, each externally licensed, each maximally economical. They are related by a derivational process that is also maximally economical. There are no superfluous symbols in representations, and no superfluous steps in derivations.

These conditions of economy of representation and derivation have a 'least effort' flavour, and thus have a kind of generality that is lacking in such principles of universal grammar as those governing referential dependence or the formation of interrogatives, discussed earlier. But this appearance is somewhat illusory. In their more precise formulation, these conditions express contingent properties of the language faculty, properties that are by no means necessary in language-like systems. Furthermore, computations are 'efficient' or 'optimal' only relative to specific design. We therefore may be saying less than we think when we describe a computation as optimal. In an apparent motion experiment, the visual system is behaving 'optimally' when it imposes continuity on discrete stimuli, even motion around a barrier rather than a straight line. The auditory system is behaving 'optimally' when it imposes discontinuity on continuously varying stimuli, as in categorical perception of /p/, /t/, and /k/, or when it interprets a continuous stimulus as a discrete sequence of words. Each system is behaving optimally relative to the way it is designed, but we learn nothing from this about efficiency in any general sense. The same is true of 'least effort'. Doubtless there is more to say, but again, caution is necessary.

Before, I mentioned that Darwin and his successors give us little reason to believe that the language faculty is designed for use or acquisition; and in fact, language design appears to be dysfunctional in certain respects, some rather basic. The properties of language design that I have now been discussing are relevant in this connection. Such properties tend to yield computational difficulties, since they impose global conditions on valid computations. In these respects, language design appears to be problematic from a parsing–theoretic perspective, though quite elegant regarded in isolation from considerations of use. With an air of paradox that is only apparent, we might say that language is 'beautiful' but not 'usable'.

The discrepancies between natural language and formal systems constructed for computational facility also suggest form–function incompatibility. Other questions are raised by such features of language design as displaced phrases, appearing in the utterance in a position different from the one in which they are interpreted. There is by now evidence from a number of languages indicating that bound empty categories are 'left behind' in the position of normal interpretation in such cases as these. Displaced phrases and empty categories also impose computational problems, but these are widespread phenomena in natural language.

Considerations of this sort are sometimes taken to be self-refuting; if a conception of language design is maladapted to some parsing model (or perhaps to *any* efficient parsing model), then that shows that the conception of language design is wrong. There are two problems with this response. The first is that independent evidence in support of parsing models is scarce, whereas the conceptions of language design in question have considerable empirical confirmation and explanatory range. A more serious problem is that the argument is based on the tacit assumption that language design is functional from a parsing–theoretic point of view and must account for the alleged fact that parsing is easy and quick. But as we have seen, parsing is not in general easy or quick, and the underlying assumption about functional adaptation has little to recommend it. The nature of language is a question of fact, not decision, and the question of how well language design is adapted to the problem of parsing is simply an empirical matter. There is no basis for the arguments that are commonly advanced in this connection. Considerations of parsing do not, by their nature, have any privileged character as regards what is called 'the psychological reality of grammar', that is, the truth of theories of I-language.

Structural properties of expressions are readily accessible in a large enough class of cases to allow for language to be usable in practice. But language design as such appears to be in part dysfunctional, yielding properties that are not well adapted to some of the functions language is called upon to perform, including communicative functions. Similar considerations arise elsewhere. Typically, we expect biological systems to have large-scale redundancy, a property that would be conducive to resistance to injury and perhaps to avoidance of computational difficulties by allowing alternative modes of operation. But it has been striking over the years to see how the contrary assumption about language appears to be valid. When proposed principles of language are redundant, in that they overlap in their consequences, it has repeatedly been discovered that they are incorrectly formulated and that the right formulation eliminates these redundancies. We must guard against the possibility that this is a feature of our research strategies, not a property of the system under investigation. But the guiding intuition that redundancies, asymmetries, and formal inelegance are signs of error has proven so productive that it has gained more than a little credibility. Language design seems to be of the sort that one expects to find, for unknown reasons, in the study of the inorganic world, but different from what is often found in other biological systems.

Possibly these features, if they are real, are related to other unusual properties of natural language, specifically, the fact that it is a digital computational system. It is not obvious how the brain could have produced such a system, and there are few known analogues in the biological world. Speculations about natural selection are no more plausible than many others;

perhaps these are simply emergent physical properties of a brain that reaches a certain level of complexity under the specific conditions of human evolution.

I should emphasise again that there are no real paradoxes here; there is no reason to take for granted that the general design of language is conducive to efficient use. Rather, what we seem to discover are some intriguing and unexpected features of language design, features that are unusual among the biological systems of the natural world.

FURTHER READING

Chomsky, N. *Knowledge of Language*, New York: Praeger, 1986.
Chomsky, N. *Language and Problems of Knowledge*, Cambridge, Mass: MIT Press, 1988.
Lasnik, H. and Uriagereka, J. *A Course in GB Syntax*, Cambridge, Mass: MIT Press, 1988.
Radford, A. *Transformational Grammar*, Cambridge University Press, 1988.
Riemsdijk, H. van and Williams, E. *Introduction to the Theory of Grammar*, Cambridge, Mass: MIT Press, 1986.

4

Telling the truth

D H MELLOR

Few witnesses in court refuse 'to tell the truth, the whole truth and nothing but the truth' on the grounds that they don't know what the truth is or how to tell it. Outside the courts, however, Pontius Pilate has been but one of many who have claimed, more or less sincerely, not to know what truth is. I can't say I know all about it either; but I do know enough to be able to tell at least some of the truth about why and how we tell, i.e. communicate, the truth. And to do that I do have to say something about what truth is, in order to explain why we should want it told. This doesn't mean I have to define truth, merely produce two important truisms about it, each of which has in fact been proposed as a definition. But for present purposes it really doesn't matter which if either of them is the right definition. All that matters is that they're both true, which I hope you'll agree they obviously are.

My first truism is the one Aristotle used to say what it is for a statement to be true or false: 'To say of what is, that it is not, or of what is not, that it is, is false; while to say of what is, that it is, or of what is not, that it is not, is true.' For example, to say of what is in fact honey that it isn't honey, or of what isn't honey that it is, is false; whereas to say of what is honey that it is honey, or of what isn't that it isn't, is true.

Now whether or not that's the right definition of truth, it is at least obviously true. And it's obviously true not only of statements, but also of beliefs, which play an absolutely crucial role, not only in telling the truth, but also in an essential preliminary to telling it: namely, finding it out. For what finding out the truth means is getting true beliefs for ourselves; and what telling it means is giving our own true beliefs to other people – often of course

Figure 1 Truth: 'To believe or say truly is to believe or say of what is that it is ...'

(though as we shall see, by no means always) by making true statements.

So I shall use Aristotle's truism in the following form, extended to cover beliefs as well as statements: *to believe or say truly is to believe or say, of what is that it is, or of what is not, that it is not.* See Figure 1.

That's my first truism about truth. The second truism is one I shall need to answer the question: why should we want to be told the truth? Or more generally, why should we want to find it out, whether by being told it or otherwise? Why, in other words, should we want to get true beliefs rather than false ones?

The reason is not a moral one. True beliefs aren't generally better than false ones in any moral sense: there is usually nothing morally wrong about being mistaken in one's beliefs about matters of fact. Sometimes there is, especially when one has to act on one's beliefs in ways that affect other people. Thus a false belief that a man is about to set off a bomb might well be reprehensible in a soldier, who therefore shoots him, when the soldier could and should have known that the man was not about to do any such thing.

But most of our mistakes have no such moral consequences. There's no moral virtue in the truth of most of our true beliefs. But there is a practical one. What is generally and inherently good about getting true beliefs is that they're useful, in the following sense: *truth is that property of our beliefs which ensures that the actions they make us perform will succeed.* That's my second truism about truth.

Take the soldier who shoots a man to prevent an explosion. What makes him do that is his belief that the man is about to explode a bomb. If his belief is true, then his action (shooting the man) will succeed: it will prevent an explosion. If it isn't, it won't: since there wouldn't have been an explosion anyway.

It's clear enough there how the truth of the soldier's belief makes the action it causes succeed. But what does it mean in general for actions caused by beliefs to succeed? To answer that, I must first say something about how beliefs cause actions. And the first thing to be said is that, on their own, beliefs don't cause actions. Believing that a man is about to set off a bomb won't make our soldier do anything, unless he also wants something: in this case, to prevent an explosion. And what the soldier wants will enormously affect what this belief of his will make him do. It's only because he wants to prevent the bomb going off that this belief of his makes him shoot. If he too had wanted the bomb to go off, his belief wouldn't have made him shoot, it would have made him dive for cover.

So what really causes the soldier's action is not just his belief, but a combination of that belief with a certain desire. And this is true of all actions: every action is caused by some combination of belief and desire. Thus suppose, to take a less bloodthirsty example, that Pooh's desire for honey makes his belief that there's some in the cupboard cause him to go to the cupboard to get it. If Pooh's belief is true, his action (going to the cupboard) will succeed: it will get him the honey he wants. But if his belief about where the honey is is false, his action will fail: it won't get him what he wants.

In short, an action succeeds when it fulfils (i.e. achieves the object of) the desire that has combined with some belief to cause that action. And that's what the truth of our beliefs ensures: that the actions they combine with our desires to cause will succeed in fulfilling those desires. That in the end is why we want true beliefs rather than false ones. We want them because truth is what makes our beliefs useful to us in this well-defined sense. Indeed, in many cases, we do more than merely want true beliefs: we positively need them in order to survive, since our survival depends on our actions fulfilling our most basic desires, such as the desire for food and warmth.

Given then that we need and want our own beliefs to be true, the next question is: how do we get them? How do we set about getting the true beliefs that we need if our actions are to succeed in fulfilling our desires? Well, obviously, we get them either for ourselves, or from other people. We get them from other people by communication; and we get them for ourselves by thinking, or by observation, or both. Beliefs that we could get just by thinking (such as beliefs about logic and mathematics) I'm going to ignore. To keep things relatively simple, I'm going to stick to true beliefs that we *could* get by observation, like beliefs about the presence or absence of honey, even though in fact we get most of them by communication: by being told things.

How then do we get true beliefs of the sort that we could get by observation? And how in particular do we get them by communication? That's my main question. But to answer it, I must make a considerable digression, to say something about how we get such beliefs by observation. And first I must say why I need to make this digression.

I'm not making it just because observation comes before communication, although it does: since before an observable truth can be communicated, someone somewhere must get that true belief – or something from which it can be inferred – by observation. Nor am I digressing because observation is a necessary part of communication, although again it obviously is: since communications obviously can't work if they're not observed. For example, I obviously can't tell you anything unless and until you hear what I've said or see what I've written.

However, the real reason for starting with observation is that communication doesn't merely depend on observation in these two ways: in effect, communication itself is a kind of observation. Being told an observable truth is, as we shall see, just one way among many of indirectly observing it. So in order to understand how observable truths are communicated, we must first understand how they are observed: first directly, and then indirectly. Only then will we be able to see what's so special about observing them by being told them.

How then do we get true beliefs by observation? Well, the simplest way, when it's feasible, is by direct observation. We just look and see, or hear, or touch or smell – or, as in Figure 2, taste.

The first thing to note about the observation that Pooh is making in Figure 2 is that it is, amongst other things, an action: something which Pooh does. Which means, as we've already remarked, that it's caused by some combination of desire and belief: in this case, by Pooh's desire, not just for honey, but to get a true belief about whether what's in the pot *is* honey, and his belief that the way to get that true belief is to taste what's in the pot.

So considered as an action, Pooh's observation will succeed if this belief of his is true: that is, if tasting what's in the pot will in fact make Pooh believe it's honey if it is honey, and believe it's not if it's not. If his tasting what's in the

Figure 2 Direct observation: How to get a true belief (1): Look and see.

pot will do that, his action will succeed: it will be a good observation. And what will make it good is a causal link, between what he's observing (what's in the pot) and the belief he gets about it. What sets up the causal link in this case is the fact that honey has a distinctive taste, which Pooh will recognise. So if what's in the pot is honey, its taste will cause Pooh to believe that it's honey. And if it isn't honey, the absence of that taste will cause Pooh to believe that it isn't honey. So either way Pooh will make a good observation. The belief he gets will be true, because he will have been caused to get it by the very fact (that what's in the pot is honey) which makes it true. And that in general is what makes direct observations good. The facts that cause the beliefs those observations yield are the very facts which make those beliefs true.

But observations needn't be direct to be good. Instead of looking directly for something one wants to get a true belief about, one can look instead for a *sign*. By 'sign', I should say, I mean nothing very technical. I mean only what anyone would mean by saying that clouds mean (i.e. are a sign of) rain or, as in Figure 3, that bees mean honey.

In Figure 3, Pooh observes the presence of honey only indirectly: what he observes directly are honey bees. And since Pooh knows what bees look like, we may assume that this observation is a good one: the belief the bees give him (namely, that they are bees) will be true. And from that belief Pooh then derives the belief that there's honey by inferring it, via his belief that bees mean honey, i.e. that where there are bees there'll be honey. And as for

Figure 3 Indirect observation: How to get a true belief (2): Look for a sign.

Pooh, so for us. We make an indirect observation by first making a direct observation of a sign, and then making an inference from that to what we believe the sign signifies.

For an indirect observation to be good, therefore, both parts of it must be good. The direct observation must be good, and so must the inference. That is, the inference must preserve the truth of its premise (that the sign is present) in its conclusion (that what the sign signifies is present). And it will do that just in case, as a matter of fact, the sign is correlated with what we take it to signify: that is, provided that, at least in Pooh's neck of the woods, there really is honey wherever there are bees. If that's so, then Pooh can get true beliefs about honey just as well by observing it indirectly by observing bees as he can by observing it directly. Either way, the beliefs he gets about honey will be true, which is what matters.

We get many of our beliefs by indirect observation in just this way, either because we don't want to make a direct observation or because we can't. We might for example want to find out if it's freezing outside without going outside to feel directly how cold it is. So we make an indirect observation by looking through the window for signs of freezing, such as frost. Or we might want to know how cold it is to the nearest °C, in which case we have no choice. We can't make that observation directly, because our feelings of cold don't enable us to discriminate temperatures that finely. This observation is one which we have to make indirectly: by directly observing a thermometer, and inferring that the temperature is what the thermometer says it is.

In both these cases, we are essentially doing just what Pooh does when he observes honey indirectly by inferring its presence from that of the bees he observes directly. There are of course obvious differences, but they're not really relevant. One such difference is that whereas thermometer readings are caused by the temperatures they signify (that's what correlates them), with bees and honey it's the other way round: what correlates them is the fact, not that honey makes bees, but that bees make honey – the sign causes what it signifies. But that's immaterial. It doesn't matter whether a sign causes what it signifies, or is caused by it, or whether both are caused by something else (like thunder and lightning, either of which might be a sign of the other and both of which are caused by an electrical discharge). What matters is the correlation between them, not how the correlation is produced.

Another and even more importantly irrelevant difference is that thermometer readings are *linguistic*: they say explicitly what they're signs of. Every reading on a good thermometer is correlated with the temperature it names: so that, for example, the reading '10°C' is correlated with that very temperature. Whereas bees, of course, like frost, are not linguistic signs: they don't say what they're signs of. But why should they? You can learn to use a sign without its having to tell you what it signifies every time you use it. Bees do not need to be labelled 'Honey' any more than honey pots do: all that's needed in each case is a memorably distinctive appearance. So that difference too is irrelevant. All we need, in order to make a good indirect observation of something, is a learnable correlation between it and something else which we can observe directly.

And that – a learnable correlation – is also the main thing we need in order to be told the truth. Suppose for example, as in Figure 4, that Pooh, visiting Rabbit, asks him if there's honey still for tea, and Rabbit says yes, there is. That statement of Rabbit's will be true (as our first truism about truth tells us) if and only if there really is honey still for tea. So in order to be true, Rabbit's statement must be correlated with what in the circumstances it says it signifies (namely honey). Otherwise the statement will be false; and so therefore will the belief, that there's honey, which Pooh gets by believing what Rabbit says.

In other words, Pooh's relation to Rabbit's statement, telling him that

Figure 4 Communication: How to get a true belief (3): Ask an informant.

there's honey, is essentially the same as his relation to the bees which he uses to observe indirectly that there's honey. Indeed from his point of view, that of the *tellee* (if you'll pardon the expression), being told the truth just *is* finding it out by a certain kind of indirect observation.

The only difference in this case is that, instead of looking out for bees, Pooh gets Rabbit to say something ('Yes') which he hears (i.e. directly observes) and understands (i.e. takes to say, as a response to his question, that it's a sign of honey). And from this Pooh infers that there really is honey. But this in essence is just what Pooh did when he inferred the same belief about honey from his direct observation of bees. In both cases what really matters is the same: that the sign he observes directly – the bees, Rabbit's saying 'yes' – should be correlated with honey, so that the belief he infers from the sign will be true.

This for Pooh is the whole object of the exercise: getting a true belief about honey. The fact that he gets it by being told it, as opposed to observing it for himself, is incidental. Communication here is a means to an end (the acquisition of a true belief), not an end in itself. And Rabbit, knowing this, could in fact have answered Pooh's question without telling him anything at all. In other words, he needn't have *told* Pooh that there was honey, he could have *shown* him that there was – for example, by drawing his attention to the pot of honey on the sideboard.

But what then is the difference between being told the truth and being shown it, and does the difference really matter? Since in both cases the end, getting a true belief, is the same, the difference must lie in the means; but what the difference is, and why it matters, remains to be seen.

The difference is this. When Pooh is shown the honey, he doesn't get his belief that there's honey from Rabbit. In fact, Rabbit needn't have that belief: he might have forgotten all about his honey until Pooh's question made him look for it. Whereas when Rabbit tells Pooh there's honey, Pooh does get his belief from Rabbit; and specifically, from what Rabbit himself believes. For Pooh doesn't infer the presence of honey directly from what Rabbit says: he infers it indirectly, via what (he believes) Rabbit believes.

In other words, what Pooh infers first from Rabbit's saying 'Yes' is that Rabbit believes there's honey. Only then, from that, does he infer that there really is honey; and this is what really distinguishes being told the truth from finding it out in other ways. When Pooh infers the presence of honey from the bees he sees, his observation may not be direct, but his inference is: bees, therefore honey. The bees' beliefs about the matter (if any) don't come into it. And when Rabbit *shows* Pooh the honey, again the inference is direct: a honey pot, therefore honey. Rabbit's beliefs about the matter don't come into it.

But when Rabbit *tells* Pooh that there's honey, Pooh's inference is indirect: it goes via Rabbit's belief. Pooh only believes what Rabbit says because he believes that Rabbit believes it too. In other words, he adopts Rabbit's belief. And that's what makes this a *communication*: the way the teller's belief is passed on – communicated – to the tellee. That's the difference that matters between Pooh believing Rabbit and Pooh believing his bees.

The other differences between these different ways of acquiring true beliefs are irrelevant. In particular, it's quite irrelevant that Rabbit tells Pooh there's honey by *saying* that there is, i.e. by producing a *linguistic* sign, whereas bees, as we've remarked, are not a linguistic sign. But as we've seen, not being linguistic doesn't make Pooh's bees any less useful as a sign. And just as indirect observation in general doesn't need linguistic signs, nor does the special case of communication. Language may not only be ill-adapted to communication (as Professor Chomsky shows us in chapter 3), it's also in principle, and quite often in practice, unnecessary. Rabbit doesn't have to use language to tell Pooh that there's honey. He needn't say anything to do that. He could just nod, or sigh, or do anything, in fact, which Pooh would rightly take to correlate with Rabbit's believing that there's honey. As indeed, in Figure 5, Eeyore does when Pooh asks him if there are any thistles.

Figure 5 Non-linguistic communication.

Eeyore's sigh is not a linguistic sign. It's like the bees: it doesn't say what it's a sign of. But it is a sign nonetheless, and what it signifies is that Eeyore believes he's out of thistles: simply because he only sighs (when the subject's raised) when he does believe that. And because Pooh has learned this correlation, just as he's learned that bees correlate with honey, Eeyore can tell Pooh that thistles are off simply by sighing, without using any language at all. In short, it's not the use of language that distinguishes being told the truth from other ways of finding it out: it's the fact that when we tell people the truth, we do so by getting them to believe what we believe.

But why do we do that? Why, for a start, do we want to be tellees, i.e. to adopt other people's beliefs? And why, as tellers, when we want to tell people the truth, do we do so by telling them instead what we believe?

The first question is relatively easy. The reason we want to adopt other people's beliefs is that we know that everyone wants their own beliefs to be true: because, as we've seen, truth is what makes our own beliefs useful to us in the way I described earlier, by making our actions succeed in fulfilling our desires.

We all know therefore that everyone tries to get their own beliefs by methods that will maximise their chances of being true: like Pooh getting his belief about what's in his honey pot by tasting it. So if, for example, I believe that you've got your belief about whether there's honey by an especially good method which I can't use (because I can't get at your honey pot), then I will naturally want to adopt your belief, in order to acquire with it its high chance of being true. In other words, the fact that we get most of our true beliefs from other people, whom we believe are better placed than we are to get them for themselves, is just a special case of the division of labour: namely, of epistemic labour, the labour of acquiring knowledge.

That's why, when I want to be told something, the belief I want to get about it is my teller's belief. There is really no great mystery about that. But that doesn't answer my second question. It doesn't explain why, when I want to tell other people the truth, I want them to believe what I believe. Because what I really want, after all, is to give them a true belief, and I know very well that, although I want my own beliefs to be true (because that's what makes them useful), we can all make mistakes. So I don't flatter myself that someone who gets my beliefs will automatically get true ones. And yet, when I want to tell people the truth, what I will in fact do is try to get them to believe what I believe. Why?

There is one very bad answer to this question, which is depressingly common and goes like this. 'We can't really know what actually goes on in the world, like whether there really is honey: all we really know, and therefore all we can really tell other people, is what we believe goes on in the world.' That's nonsense. We know far more about what goes on in the world, i.e. we have far more reliably formed true beliefs about it, than we do about our own beliefs about what goes on in the world: beliefs about honey, thistles, etc. We have very few beliefs, true or false, about what our beliefs about honey etc. are: why should we? At any one time, therefore, most of the many beliefs that constitute our knowledge of what goes on in the world are beliefs that we don't know we have. So it's just not true that we know less about what's going on than we know about our own beliefs about what's going on. It's absolutely the other way round. So that can't be why, when we want to tell people the truth about what's going on, what we actually do is try to give them our own beliefs about what's going on.

No, the real reason is this. Telling the truth about something is an action, caused, like all actions, by a combination of desire and belief: in this case, of a desire to tell some truth and a belief about what the relevant truth is. So suppose Rabbit wants to tell Pooh the truth about honey, and believes the relevant truth to be the proposition, P, that there is some. But by our first truism about truth, for it to be true that there's honey is just for there to be honey. So for Rabbit to believe that P is true is just for him to believe P, i.e. to believe that there's honey. That's why his desire to tell Pooh the truth will in fact make him tell Pooh what he believes, whether that is actually true or not.

This is almost right, but not quite. Telling the truth is a little more complicated than this. For Rabbit doesn't just want to *say* what's true: he wants to make Pooh believe it. And as an experienced tellee himself, he knows that Pooh will only believe what he says if Pooh believes that he believes it too. So Rabbit's *immediate* desire is to give Pooh a true belief about what he, Rabbit, believes. So what Rabbit will tell Pooh is not necessarily what he actually believes, but what he *believes* he believes. But since (Rabbit believes) Pooh will in fact believe that Rabbit believes what he says, this needn't make Rabbit say 'I believe there's honey': it need only make him say 'There's honey'.

And as for Rabbit, so for the rest of us. When we set out to tell other people the truth by saying things, what we actually do is to say not what we believe, but what we believe we believe – which, as Freud and others have taught us, isn't always the same thing.

That – I'm telling you! – is how we tell the truth. But I fear you may not believe me: because you may well think that I've made the process seem incredibly complicated. After all, we all know how often we tell the truth, and I'm sure it doesn't seem anything like as complicated a process as I've said it is.

Nor it does. But it is. The reason the process of telling the truth seems less complicated than I've said is simply that we aren't conscious of most of the mental processes I've been describing. But one of the most persistent and pernicious myths we've inherited from Descartes is that mentality is essentially conscious, so that anything we can't introspect can't really be going on in our mind. But we know now that that's not true: that there are many unconscious and subconscious mental processes which we can't just introspect, and that our mental life is far more complicated than we ourselves are ever aware of at the time.

So what I've said should not be incredible in principle. And in practice, I can even offer you an introspectible piece of evidence for it, as follows. I've said that most of the time we don't know what our own beliefs are, because we don't believe we have them. (By which, of course, I don't mean that we *dis*believe that we have them: merely that we mostly have no belief either way about what beliefs we have.) So when, for example, Pooh goes to get honey from the cupboard where he believes it is, he needn't be aware of having that belief. He can just go, guided by that belief, which he has, but which he needn't at that instant believe he has.

But if I'm right about what it takes to tell the truth, Pooh can't *tell* anyone that there's honey there without first becoming aware of having that belief. And that's because, according to me, telling the truth means saying (or otherwise conveying) not what you believe, but what you believe you believe. And similarly of course when you want to lie or to mislead – to give someone a false belief – what you'll say is not necessarily something you disbelieve, but rather something you believe you disbelieve. So either way, whether you want to tell the truth or to lie, you need to have beliefs about what your relevant beliefs are. In other words, you have to be aware of them.

And so you do. Wanting to tell people things, sincerely or not, does demand an awareness of the beliefs (or disbeliefs) you're trying to give them: an awareness which most of the actions those beliefs combine with your desires to cause doesn't demand at all. And this is just a fact: indeed an introspectible fact, which we can therefore all observe directly for ourselves. So this is not something I'm even trying to tell you: it's something I'm trying to *show* you, by drawing your attention to it! All I'm trying to *tell* you is how my account of how we tell the truth explains this fact, which apparently simpler accounts of how we tell the truth don't do. And that fact, I believe, provides significant support for my account.

So much for telling the truth. What about lying? Suppose Rabbit doesn't in fact believe there's any honey left (and believes he doesn't believe that). So when he says 'There's honey', he's lying, saying something he believes to be false. But now suppose that Eeyore does the same. Suppose he sighs because, although (he believes) he believes there are thistles left, he wants to keep them to himself. Has he lied to Pooh, or just misled him? He hasn't after all said anything false, because he hasn't said anything at all: all he's done is sigh. And some people think that this matters: that when it's wrong to mislead people – which it usually, if not always, is – then it's not quite as bad if you can manage to do it without actually *saying* anything you believe to be false.

I think that's nonsense. The only thing that's ever wrong with saying something you believe to be false is that you do it in order to mislead someone whom you think will believe what you say. There is after all nothing inherently wrong with quoting fiction – with saying, for example, that Baker Street once housed a detective called Sherlock Holmes – so long as you don't mislead anyone by palming it off as a fact. And if you did palm it off as a fact, it would be no excuse that you had done it non-linguistically: for example, by including clips from Sherlock Holmes movies in old newsreels as if they were genuine news items. What matters about lying is giving people false beliefs, just as giving them true ones is what matters about telling the truth. Whether that's done by saying things or not is immaterial. So I'd say that Eeyore too was lying, or at least doing something just as bad.

But how can you tell when other people are lying? Well, sometimes it's easy, because you know independently, not only whether what they say is true, but whether they believe it. Suppose for instance you see some visibly sighted person make a phone call in broad daylight, and hear them say that it's pitch dark. You know they're lying because you can see, not only that it isn't pitch dark, but that anyone who isn't blind can see that too.

Of course it isn't always as easy as that. And when it isn't, one maxim that's often used as a lie detector is the maxim that *actions speak louder than words*. And I'd like to end by saying why that's often (though not always) true.

One might, for instance, use the maxim to infer that because Eeyore's sigh is a non-linguistic action, it's a better sign that he's out of thistles than Rabbit's words are that he has some honey. But even if that's true, it won't be because sighs aren't words. It will be because Eeyore isn't even trying to communicate. For example, his sigh might not be a voluntary action at all. It might be an involuntary reaction, which any mention of thistles always produces in Eeyore when he believes he's out of them. And if it is, then its correlation with that belief of his won't depend on his wanting anything (other than thistles) and in particular not on his wanting to tell anyone the truth. And this will make Pooh's inference from Eeyore's sigh to what Eeyore believes safer than his inference from Rabbit's words to what Rabbit believes: because that inference does depend on Rabbit's wanting to tell Pooh the truth, whereas, as we've remarked, he may in fact be lying.

But then, as we've also remarked, so may Eeyore be lying. He too may be trying to mislead Pooh, by sighing deliberately when (he believes) he believes he's by no means out of thistles. So the point of the maxim that actions speak louder than words is not that people never use non-linguistic actions to communicate (which is when they may be deliberately misleading), but that language is much less often used to do anything else.

So while you might well overhear Eeyore sighing to himself, and know therefore that he isn't trying to mislead anyone, it's much less likely (though not of course impossible) that you'll overhear Rabbit muttering 'There's honey' to himself. That's why actions generally, although by no means always, do speak louder than words: because, paradoxically, they aren't meant to 'speak' at all, and *a fortiori* aren't meant to speak what's false.

That's all I have to tell you about how to tell the truth. Or rather, about how to *try* to tell it: since whether any such action of yours succeeds in doing what you want (giving your tellee a true belief) will depend as we've seen on the truth of the beliefs which you also need in order to make you undertake that action. And I don't just mean the belief that you're trying to communicate. There's also your belief about what that belief of yours is, which is what will determine what you'll actually say. And finally, of course, there's your belief that your tellee will believe what you say. If all those beliefs of yours are true, your action will succeed. You will give your tellee a true belief: you will actually tell the truth. If they're not all true, then you'll probably fail. But at least you'll have tried. As I have done.

5

The novel as communication

DAVID LODGE

Anyone contributing a single lecture to a series like this – covering a wide spectrum of disciplines and addressed to a general audience – is bound to feel some anxiety and uncertainty about how their contribution will fit into its larger context. Looking at the other titles I infer that mine is the only lecture on literature, and logically, perhaps, my title should be '*Literature* as communication'. Some of what I have to say does in fact apply to literature generally, but much of it is specific to the novel, the literary form in which I am most interested, both as a literary critic and as a practising writer.

There are two possible ways of approaching this topic. One is to take for granted that the novel is a mode of communication, and to analyse its formal features as techniques of communication; the other is to question the assumption that the novel is communication – to ask what is implied by that assumption, and what excluded. I shall try to do a little of both. I shall also consider the subject from two points of view: that of a critic and that of a practising novelist.

I suspect that it is assumed by most people, including those who planned this course of lectures, that language is a means of communication – that this is what it is *for*; and that since literature is made out of language, it too must be a kind of communication, as defined by, for instance, the Collins English Dictionary: 'the imparting or exchange of information, ideas, feelings'. A commonsense view of the matter would say that that definition covers the composition and reception of a novel.

The classic novelists certainly seem to have thought of their activity as communication. Henry Fielding, for example, in the eighteenth century,

The novel as communication: *La Liseuse de roman* by Vincent van Gogh

draws his masterpiece, *Tom Jones*, to its conclusion with a metaphor of social intercourse:

> We are now, reader, arrived at the last stage of our journey. As we have, therefore, travelled together through so many pages, let us behave to one another like fellow-travellers in a stage coach, who have passed several days in the company of each other; and who, notwithstanding any bickerings or little animosities which may have occurred on the road, generally make up at last, and mount for the last time into their vehicle with cheerfulness and good humour; since after this one stage, it may possibly happen to us, as it commonly happens to them, never to meet more.

The intrusive authorial voice exemplified in this passage, and generally typical of the classic novel – the voice that confides, comments, explains and sometimes scolds – the voice to which we rather casually give the name that appears on the title-page (Henry Fielding, Charles Dickens, George Eliot, or whoever) is the most obvious sign that these writers saw themselves as engaged in an act of communication with their readers. In this kind of novel the act of narration is modelled on a speech act in which one person tells a story to another. George Eliot begins her novel *Adam Bede* thus:

> With a single drop of ink for a mirror, the Egyptian sorcerer undertook to reveal to any chance comer far-reaching visions of the past. This is what I undertake to do for you, reader. With this drop of ink at the end of my pen, I will show you the roomy workshop of Jonathan Burge, carpenter and builder in the village of Hayslope, as it appeared on the 18th of June, in the year of Our Lord, 1799.

By apostrophising the reader, the act of writing is transformed here into a kind of speaking. Through the figure of the drop of ink, at once miraculous and homely, the act of telling is transformed into a gesture of showing. This offer to transport us out of our own world, with all its problems, unfinished business, boredom and disappointment, into another world where we may escape these things or negotiate them vicariously, is perhaps the fundamental appeal of all narrative. What is peculiarly novelistic about George Eliot's opening gambit is its pseudo-documentary specificity – the proper names and the date: 'the roomy workshop of Jonathan Burge, carpenter and builder in the village of Hayslope, as it appeared on the 18th of June, in the year of Our Lord, 1799'.

The novel is a form of narrative. We can hardly begin to discuss a novel without summarising or assuming a knowledge of its story or plot; which is not to say that the story or plot is the only or even the main reason for our interest in a novel, but that this is the fundamental principle of its structure. (These two terms, incidentally, story and plot, are sometimes used to describe two opposed types or aspects of narrative, but unfortunately they are also used as interchangeable synonyms and I shall use them as such.) The novel therefore has a family resemblance to other narrative forms, both the purely verbal, such as the classical epic, the books of the Bible, history and biography, folktales and ballads; and those forms which have non-verbal components, such as drama and film. Narrative is concerned with *process*, that is to say, with change in a given state of affairs; or it converts problems and contradictions in human experience into process in order to understand or cope with them. Narrative obtains and holds the interest of its audience by raising questions in their minds about the process it describes and delaying the answers to these questions. When a question is answered in a way that is both unexpected and plausible, we have the effect known since as *peripeteia* or reversal. All this applies to the novel as to every other form of narrative, whatever its medium.

Verbal narrative, as distinct from narrative which includes an element of performance and visual images, has two basic modes of representation: the report of characters' actions by a narrator, and the presentation of the characters' own speech in dialogue. These two modes – narrator's voice and character's voices, or summary and scene, telling and showing, as they are sometimes called – are the woof and warp of all verbal narrative, from the story of Little Red Riding Hood to *War and Peace*. The novel, however, exhibits particularly subtle and complex interweavings of these modes of presentation, as I shall indicate more fully later on. Its discursive variety and complexity is one of the reasons why it imitates the social world with a verisimilitude unequalled by other literary forms. Another reason is that pseudo-documentary specificity I mentioned earlier in connection with the opening of George Eliot's *Adam Bede*. In short, the novel is characteristically a *realistic* form of narrative. Early critical discussion of the novel, which acquired a distinctive generic identity in the eighteenth century, focussed on this quality: its illusion of reality. Clara Reeve, for instance, writing in 1785, said:

> The Novel gives a familiar relation of such things as pass every day before our eyes, such as may happen to our friends, or to ourselves, and the perfection of it is, to represent every scene in so easy and natural a manner, and to make them appear so probable, as to deceive us into a persuasion (at least while we are reading) that all is real, until we are affected by the joys or distresses, of the persons in the story, as if they were our own. (*The Progress of Romance*)

A much later, much more sophisticated critic, Ortega y Gasset, said much the same thing:

> ...the novel is destined to be perceived from within itself – the same as the real world...to enjoy a novel we must feel surrounded by it on all sides... Precisely because it is a preeminently realistic genre it is incompatible with outer reality. In order to establish its own inner world it must dislodge and abolish the surrounding one. ('Notes on the novel', 1948)

Out of a thousand possible illustrations of this point one might cite the testimony of William Smith the nineteenth-century publisher, on his first reading of the manuscript of Charlotte Brontë's *Jane Eyre*.

> After breakfast on Sunday morning I took the ms of Jane Eyre to my little study and began to read it. The story quickly took me captive. Before 12 o'clock my horse came to the door, but I couldn't put the book

> down. I scribbled 2 or 3 lines to my friend saying I was sorry circumstances had arisen to prevent my meeting him, sent the note off to my groom, and went on reading the ms. Presently the servant came to tell me lunch was ready. I asked him to bring me a sandwich and a glass of wine, and still went on reading Jane Eyre. Dinner came; for me the meal was a very hasty one, and before I went to bed that night I had finished reading the manuscript.

The peculiar verisimilitude of the novel's representation of reality, and the peculiarly hypnotic spell the novel casts upon its readers, have always made it an object of some suspicion, both morally and aesthetically. Is there not something fundamentally unnatural and unhealthy about a form of art which suspends the reader's awareness of his own existence in real space and time? Is not the pleasure of the novelistic text akin to day-dreaming, wish-fulfilment fantasy? Freud certainly thought so (see his paper on 'Creative writers and day-dreaming'). On such grounds it has been argued that the novel is not authentic communication, notably by the Marxist critic Walter Benjamin.

Benjamin drew a distinction between storytelling, which he saw as, in its purest form, an oral–aural transaction between a narrator and an audience physically present to each other, and the novel, which is produced in one place by a solitary silent author, and consumed in another place by a solitary silent reader. The rise of the novel, he observed, was coincident with the decline of storytelling; and in consequence he says, in a striking phrase, 'the communicability of experience is declining'.

> The novelist has isolated himself. The birthplace of the novel is the solitary individual, who is no longer able to express himself by giving examples of his most important concerns, is himself uncounselled and cannot counsel others. ('The Storyteller')

In recent years there have been many attacks on what is sometimes called the classic realist novel on similar grounds: that far from being a means of communication it is a means of ideological domination and repression, reproducing on the cultural level the processes of industrial capitalism, making its audience passive consumers, reconciling them to their alienated state instead of liberating them from it, by making it appear normal or natural. (One of the most recent of such polemics is *Resisting Novels: Ideology and Fiction*, by Lennard J. Davis, 1987).

In fact the classic novelists were well aware of the dangerous power of their art, and took various measures to prevent or warn against its abuse.

The intrusive authorial voice itself is often used to point to the formal conventions of the novel, and thus to prevent a naive confusion of literature with life. When Henry Fielding introduces the word 'pages' into his stage coach metaphor ('As we have, therefore, travelled together through so many pages') he reminds his audience that they are reading a book. Jane Austen gives an even sharper jolt at the end of *Northanger Abbey* to 'my readers, who will see in the tell-tale compression of the pages before them, that we are all hastening together to perfect felicity'. When Trollope says in *Barchester Towers*, 'But let the gentle-hearted reader be under no apprehension whatsoever. It is not destined that Eleanor shall marry Bertie Stanhope', he is teasing rather than indulging his audience.

Henry James did not see it that way. Such authorial admissions that the events of the novel are invented seemed to him 'a betrayal of a sacred office'. James inaugurated the modern or as it is sometimes called, the 'modernist' novel in England, a kind of fiction which, in pursuit of a more faithful representation of reality, attenuated or eliminated altogether the authorial narrator. Instead the action is narrated as perceived by the consciousness of a character or characters. This can be done in various ways, most of which can be found in eighteenth- and early nineteenth-century fiction, but not used so artfully or extensively. One simple and obvious way of eliminating the authorial voice and giving a realistic effect to the novel is to make a character the narrator, as in *Robinson Crusoe* or *Jane Eyre* or *David Copperfield*. But whereas those first person narrators are fairly transparent surrogates for the implied authors of those novels, the first-person narrators of modernist texts are more ambiguous, less reliable witnesses to their own experience, and are often framed by or counterpointed with other narrators – as, for example, in Henry James's *The Turn of the Screw* or Conrad's *Heart of Darkness*. A further, characteristically modern variation on the pseudo-autobiographical or confessional novel is the interior monologue, as used in James Joyce's *Ulysses*, where the reader eavesdrops, as it were, on the actual thoughts and sensations of the character as he or she moves through time and space. Here is Joyce's Leopold Bloom, considering his cat:

> They call them stupid. They understand what we say better than we understand them. She understands all she wants to. Vindictive too. Cruel. Her nature. Curious mice never squeal. Seem to like it. Wonder what I look like to her. Height of a tower? No, she can jump me.

This kind of discourse is at the opposite pole from storytelling as defined by

Benjamin. Compared to the classic novel, not a great deal happens in the stream-of-consciousness novel – or it happens off-stage, as it were, glanced at in memory and allusion rather than presented directly. The minute registering of 'the flickerings of that innermost flame which flashes its messages through the brain', to use Virginia Woolf's words, works best on ordinary experience rather than extraordinary – walking along a street, preparing a meal, knitting a stocking. Interior monologue is particularly ill-suited to narrative purposes – even Joyce uses it only intermittently in *Ulysses*, in combination with other kinds of discourse, including free indirect style.

Free indirect style is a mode of narration which as it were fuses and interweaves the authorial narrator's speech and the speech of the character. By reporting the character's thoughts in the third person, past tense, as in traditional narrative, but keeping to vocabulary appropriate to the character, and omitting some or all of the tags that normally introduce reported speech (like 'he thought', 'she wondered', etc.) an effect of intimate access to the character's inner self is produced, without relinquishing the task of narrating to the character entirely, as in the pseudo-autobiography or interior monologue. This type of discourse – free indirect speech or free indirect style – is peculiar to the novel; it makes its appearance in the late eighteenth century and Jane Austen was probably the first novelist to realise its full potential. Here for instance is her rendering of Emma Woodhouse's thoughts, having just received a most unwelcome proposal of marriage from Mr Elton whom she had supposed to be in love with her protégée, Harriet, as a result of her own matchmaking contrivances:

> She had taken up the idea, she supposed, and made everything bend to it. His manners, however, must have been unmarked, wavering, dubious, or she could not have been so misled.

So far there is a discreet element of authorial summary in this representation of the heroine's thoughts as she reviews the supposed courtship of Harriet by Mr Elton. But in the following sentences – unfinished, fragmentary, as spontaneous as speech – we seem to be placed right inside Emma's head.

> The picture! – How eager he had been about the picture! – and the charade! – and an hundred other circumstances; – how clearly they had seemed to point at Harriet. To be sure the charade, with its 'ready wit' – but then, the 'soft eyes' – indeed, it suited neither; it was a jumble without taste or truth. Who could have seen through such thick-headed nonsense?

Jane Austen uses this technique sparingly, to represent moments of inner crisis, in combination with the more traditional modes of authorial report and direct speech. But Virginia Woolf's mature novels consist almost entirely of long passages of introspection by the characters in free indirect style, punctuated by banal conversational remarks and parenthetical reports of trivial actions. Here is another fictional matchmaker, Mrs Ramsay, in *To the Lighthouse* (1927):

> Foolishly, she had set them opposite each other. That should be remedied tomorrow. If it were fine, they should go for a picnic. Everything seemed possible. Everything seemed right. Just now (but this cannot last she thought), dissociating herself from the moment while she talked about boots) just now she had reached security; she hovered like a hawk suspended; like a flag floated in an element of joy which filled every nerve of her body fully and sweetly, not noisily, solemnly rather, for it arose, she thought, looking at them all eating there, from husband and children and friends...

Thus Mrs Ramsay at her dinner table, thinking her thoughts, 'dissociated from the moment', plotting a match between Lily Briscoe and William Bankes, while Lily Briscoe at the same table is thinking wistfully of quite another man.

The emergence of the stream-of-consciousness novel at the end of the nineteenth and beginning of the twentieth centuries was obviously related to a huge epistemological shift in culture at large, from locating reality in the objective world of actions and things as perceived by common sense, to locating it in the minds of individual thinking subjects, each of whom constructs their own reality, and has difficulty in matching it with the reality constructed by others. If the modern novel is a form of communication, then paradoxically what it often communicates is the difficulty or impossibility of communication. One of the modernist arguments for removing the intrusive authorial voice – wise, omniscient, reliable, reassuring – from the novel was that it was false to our experience that life is in fact fragmented, chaotic, incomprehensible, absurd. The trouble with the classic realist novel, in this view, was that it was not realistic enough: truth to life was sacrificed to the observance of purely narrative conventions. 'If a writer could...base his work upon his own feeling and not upon convention', said Virginia Woolf, in her celebrated essay, 'Modern fiction', 'there would be no plot, no comedy, no tragedy, no love interest or catastrophe in the accepted style...'. Instead she called for a kind of fiction that would record the atoms of experience 'as they fall upon the mind, in the order in which they fall', that would 'trace the pattern, however disconnected and incoherent in appearance, which each sight or incident scores upon the consciousness'.

The modernist novel thus tends to endorse the philosophical argument known as solipsism – that the only thing I can be sure exists is myself as a thinking subject. This lays it open to Walter Benjamin's critique even more than the classic realist novel, for it is still further removed from his concept of storytelling. Another Marxist critic, the Hungarian Georg Lukács, attacked modernist fiction on similar grounds. 'Man, for these writers [he said], is by nature solitary, asocial, unable to enter into relationships with other human beings.' Unable therefore to communicate, and unable to act upon history. And of course a familiar populist complaint about modernist fiction is that it does not communicate its meaning to the reader in a clear and comprehensible way. It is obscure, difficult, esoteric, elitist.

The standard defence of the modernist novel is based precisely upon these qualities, on its formal complexity and difficulty: the 'revolution of the word' is seen as either essential to, or more important than, any political revolution. The paradigmatic case is James Joyce. After the psychological hyper-realism of the early chapters of *Ulysses*, the text is taken over by a bewildering variety of voices and discourses – parodic, travestying, colloquial, literary: newspaper headlines, oratory, women's magazines, pub talk, operatic songs, encyclopaedia articles, and so on; while the narrative level of the text is full of gaps, non sequiturs, anticlimaxes, and unsolvable enigmas, and the chronological order of events is broken down and rearranged by the operations of memory and the association of ideas in the consciousness of characters. Reading such a text we are reminded that the world we inhabit is constructed, not given; constructed in language. As Gabriel Josipovici has said, 'To imagine, like the traditional novelist, that one's work is an image of the real world, to imagine that one can communicate directly to the reader what it is that one uniquely feels, that is to fall into the real solipsism, which is, to paraphrase Kierkegaard on despair, not to know that one is in a state of solipsism' (*The World and the Book*).

That is, I think, a somewhat tendentious description of the classic realist novel, and, in fact, writers like E. M. Forster, D. H. Lawrence, Ernest Hemingway, Evelyn Waugh and Graham Greene have written fiction that answers to the twentieth century's sense of moral and philosophical crisis without deviating violently from the conventions of classic realism. In the experimental fiction of our day that is sometimes called 'post-modernist' these conventions – such as the omniscient and intrusive authorial narrator – are retained in exaggerated and parodic forms that remind one of the meta-fictional jokes of Fielding, Sterne, Thackeray and Trollope (one thinks for instance of Muriel Spark and John Fowles in this respect). In short, I am suggesting that there is more continuity than discontinuity in the development of the novel as a literary form.

The emergence of the modernist novel of consciousness is often described in terms of a shift of emphasis from 'telling' to 'showing' – but showing in this context is a metaphor. Written language cannot literally show us anything except writing. Speech cannot show us anything except speech. Language is not an iconic sign system, in which the signifier has a visual resemblance to the signified (as in the traffic signs for 'falling rocks' or 'humped-backed bridge'), but a symbolic one in which the connection between signified and signifier is arbitrary. The stream-of-consciousness novel only 'shows' us the operations of the mind by another kind of telling than straightforward authorial report. And even those modern fictional texts, such as the novels of Samuel Beckett, that seem dedicated to demonstrating the impossibility of communicating anything to anybody about anything, do so by alluding to a paradigmatic act of storytelling, a paradise lost of communication:

> Where now? Who now? When now? Unquestioning. I, say I. Unbelieving. Questions, hypotheses, call them that. Keep going, going on, call that going, call that on. Can it be that one day, off it goes on, that one day I simply stayed in, in where, instead of going out, in the old way, out to spend day and night as far away as possible, it wasn't far. Perhaps that is how it began. [*The Unnamable*]

To recapitulate: the novel tells a story, which has some kind of generalisable thematic significance, by means of a tissue of interwoven discourses. There is the discourse of the narrator, who may be a character or an authorial persona, who, if the latter, may be covert or overt; and there are the discourses of the represented characters, as manifested in their direct speech, or what we usually call 'dialogue', and as manifested in the representation of their thoughts through soliloquy, reported speech, free indirect style, interior monologue and so on. But all these discourses will also contain echoes of, allusions to, anticipations of, other discourses both spoken and written – the discourses of popular wisdom, literary tradition, cultural institutions, social classes, and so on. It is this multivocal quality that distinguishes prose fiction from poetry, as Mikhail Bakhtin, the great Russian theorist whose work has only recently become well-known in the West, observed:

> The possibility of employing on the plane of a single work discourses of various types, with all their expressive capacities intact, without reducing them to a single common denominator – this is one of the fundamental characteristics of prose. Herein lies the profound distinction between prose style and poetic style... For the prose artist the world is full of other people's words, among which he must orient himself, and whose speech characteristics he must be able to perceive with a very keen ear. He must introduce them into the plane of his own discourse, but in such a way that this plane is not destroyed. He works with a very rich palette.

But this richness and complexity of discursive texture in the novel, what Bakhtin called the novel's 'polyphony', offers a certain resistance to the idea of the novel as communication.

The basic model of communication is a linear sequence:

addresser → message → addressee

The addresser encodes a message in language and sends it to the addressee via speech or writing and the addressee decodes it. But who is the addresser in prose fiction? The French critic Roland Barthes quotes a passage from Balzac's story 'Sarrasine', in which a young sculptor falls in love with a castrato disguised as a woman. The words of the text are as follows: '*This was woman herself, with her sudden fears, her irrational whims, her instinctive worries, her impetuous boldness, her fussings and her delicious sensibility.*' Barthes asks:

> Who is speaking thus? Is it the hero of the story bent on remaining ignorant of the castrato hidden beneath the woman? Is it Balzac the individual, furnished by his personal experience with a philosophy of Woman? Is it Balzac the author professing 'literary' ideas on femininity? Is it universal wisdom? Romantic psychology? We shall never know, for the good reason that writing is the destruction of every voice, of every point of origin. Writing is that neutral, composite, oblique space where our subject slips away, the negative where all identity is lost, starting with the very identity of the body writing. ['The death of the author']

It is time to consider, very briefly, the assault mounted by post-structuralist literary theory on the idea of literature as communication – indeed on the idea of communication itself.

The trouble with the model of communication in which the addresser encodes a message and sends it to the addressee, who decodes it, is, as another post-structuralist theorist has pointed out, that 'every decoding is another encoding'. Perhaps I may be permitted to quote from Morris Zapp's lecture on 'Textuality as striptease' in my novel, *Small World*:

> If you say something to me I check that I have understood your message by saying it back to you in my own words, for if I repeat your own words exactly you will doubt whether I have really understood you. But if I use *my* words it follows that I have changed *your* meaning, however slightly... Conversation is like playing tennis with a ball made of Krazy Putty, that keeps coming back over the net in a different shape. Reading of course is different from conversation. It is more passive in the sense that we can't interact with a text, we can't affect the development of the text by our own words, since the text's words are already given. That is what perhaps encourages the quest for interpretation. If the words are fixed once and for all, on the page, may not their meaning be fixed also? Not so, because the same axiom, every decoding is another encoding, applies to literary criticism even more stringently than it does to ordinary spoken discourse. In ordinary spoken discourse the endless cycle of encoding–decoding–encoding may be terminated by an action, as when for instance I say, 'The door is open' and you say, 'Do you mean you would like me to shut it?' and I say, 'If you don't mind', and you shut the door, we may be satisfied that at a certain level my meaning has been understood. But if the literary text says, 'The door was open' I cannot ask the text what it means by saying that the door was open, I can only speculate about the significance of that door – opened by what agency, leading to what discovery, mystery, goal?

In other words, the fact that the author is absent when his message is received, unavailable for interrogation, lays the message, or text, open to multiple, indeed infinite, interpretation. And this in turn undermines the concept of literary texts as communications. If Jane Austen's *Emma*, for instance, is a communication, what is its message? Hundreds of articles and chapters of books have been published, purporting to explain what that novel 'means', what it is 'about', and we can be sure that hundreds more will be published in the future. They all differ to a greater or less extent from each other in their conclusions and emphases; indeed if they did not differ there would be no need for more than one to be published. Does this mean that the message hasn't got across, that Jane Austen has somehow failed to communicate? This is the perennial paradox in which literary criticism finds itself implicated: that, as Michel Foucault observed:

> the commentary must say for the first time what had, nonetheless, already been said [by the original text] and must tirelessly repeat what had never been said [by other commentators]. Commentary...allows us to say something other than the text itself, but on condition that it is the text itself which is said, and in a sense completed. ['The order of discourse']

The fact that we cannot identify the author of a text simply and straightforwardly with any of the discourses which make it up, especially in the polyphonic novel-text, and the fact that literary texts resist interpretive closure, has led some modern critics to deny that literature is communication. Rather they see it as *production* – the production of meaning by the text itself when activated by the reader. Roland Barthes again:

> The text is a productivity. This does not mean that it is the product of labour (such as could be required by a technique of narration and the mastery of style) but the very theatre of a production where the producer and reader of the text meet: the text 'works', at each moment and from whatever side one takes it. Even when written (fixed) it does not stop working, maintaining a process of production. The text works what? Language. It deconstructs the language of communication, representation, or expression (when the individual or collective subject may have the illusion that he is imitating something or expressing himself) and reconstructs another language, voluminous, having neither bottom nor surface...

This is a forceful attack on not only the idea of literature as communication but also on the idea of communication itself. Note that communication is described as an 'illusion' which literary language 'deconstructs'. Barthes is here influenced, no doubt, by the founder of deconstruction, Jacques Derrida, who argued that contrary to the traditional view that speech is the exemplary case of language in use, and writing an artificial substitute for speech, writing ought to be privileged because it exposes the fallacious metaphysics of presence, of the autonomy of the subject, which speech encourages.

Deconstruction marginalises the author, or seeks to do away with the author altogether, replacing him or her with what Foucault called the 'author-fiction', that is, a culturally and historically determined role over which the individual writer has no control. As we see from the passage I just quoted from Barthes, the work or labour that the writer puts into composing his text is brushed aside as of no importance. Rather it is the text that 'works', and the text is not something that the author creates and hands over to the reader, but that the reader produces in the act of reading it – and by writing his own text about it. For the production-model of the literary text is a very academic one. It has its origin in the academic institution's need to justify the endless multiplication of commentaries, from undergraduate essays to doctoral dissertations and scholarly articles. It offers an escape from the double bind of commentary pithily summarised by Foucault, in the passage I quoted just now. In this perspective there is no essential distinction between primary and secondary texts, between so-called creative and critical writing.

Most writers and readers of fiction outside the academy, it must be said, still subscribe to the communication model of the literary text. That is, they regard a novel as the creation of a particular human being, who has a particular vision of the world, which he tries to communicate to his or her readers by employing the codes of narrative and language in a particular way, and is responsible for the novel's success or failure in this regard, and deserves praise or blame accordingly. That is the basis on which most novels, including my own, are actually written, published and received in our culture.

As a practising novelist, my instinctive reaction is to repudiate the deconstructionist position. Barthes says the text is not 'the product of a labour (such as could be required by a technique of narrative or a mastery of style)'. I know empirically that a novel *is* the product of such labour. (So, I believe, were Barthes's own books, if we substitute 'argument' for 'narrative' in his formulation.) But is it a labour of communication? A major difficulty, here, is that the idea of communication is tied up with the idea of intention, and intention is a very tricky concept in literary criticism. It is fairly easy to demonstrate that the meaning of a text cannot be constrained by reference to a writer's intentions. Let me give a trivial but I hope interesting example from my own experience. In *Small World* the middle-aged English academic Philip Swallow has a wife called Hilary and has a passionate affair with a younger woman called Joy, who reminds him of his wife when she was younger and prettier – when he first meets her, Joy is even wearing a dressing gown like one Hilary used to wear. Reviewing the novel in the London *Times*, A. S. Byatt noted approvingly that this theme of identity and difference was neatly encapsulated in the names of the two women, Hilary being derived from the Latin *hilaritas*, or Joy. Now I can be quite sure I had not intended this pleasing symmetry. I called Philip's wife Hilary in a previous novel, *Changing Places*, because it is an androgynous name and at that stage of their marriage she was the dominant partner in the marriage, or, as the saying is, wore the trousers. I called Joy, Joy because when Philip falls in love with her he is in pursuit of what he calls 'intensity of experience', an essentially Romantic quest with a capital R, and joy is a key word in Romanticism. At the moment of consummation, Philip shouts aloud the word 'Joy', which is both exclamation and apostrophe. I had no conscious awareness of the Latin root of the name Hilary until Antonia Byatt pointed it out to me. Nonetheless the play on words is there in the text, and is appropriate. It seems a good case of what Barthes calls the text working.

Another difficulty with the idea of the novel as an intentional act of communication is that until the writer has completed it he doesn't know what it is that he is communicating, and perhaps doesn't know even then. You discover what it is you have to say in the process of saying it. However carefully and thoroughly you prepare the ground, you cannot possibly hold the whole complex totality of a novel in your head in all its detail at any one moment. You work your way through it word by word, sentence by sentence, paragraph by paragraph, trying to hold in your head some idea of the totality to which these bits are contributing. What you have written already and what you plan to write in the future are always open to revision, though such possible revisions will be constrained by their mutual effect on each other. The future of a novel in the process of composition is always vague, provisional, unpredictable – if it were not so, the labour of writing it would be too tedious to bear. When you have finished the novel it is not that you have really finished it, but that you have decided to do no more work on it. If you sat down and made another fair copy of the manuscript, you would infallibly find yourself making new adjustments and emendations to it. And when the novel is published and goes out of your control to modify it, it also goes out of your control to intend the meaning of it. It is read by different readers in a bewildering variety of ways, as reviews and readers' letters attest. Can this be described as a process of communication?

I think it can, as long as we realise the inadequacy of the simple linguistic model of communication (addresser–message–addressee) not only to literary discourse, but to any discourse. The model only works at the level of the textbook example, the single isolated sentence. But there are no isolated sentences in reality. Here we must reintroduce Bahktin. Language, according to Bakhtin, is essentially dialogic. Everything we say or write is connected both with things which have been said or written in the past, and with things which may be said or written in response to it in the future. The words we use come to us already imprinted with the meanings, intentions and accents of others, our speech is a tissue of citations and echoes and allusions; and every utterance we make is directed towards some real or hypothetical Other who will receive it. 'The word in living conversation', says Bakhtin, 'is directly, blatantly directed towards a future answer word. It provokes an answer, anticipates it and structures itself in the answer's direction.'

The same is true of literary discourse, in a more complicated way. To write a novel is to manipulate several different codes at once – not simply the linguistic codes of grammar and lexis, denotation and connotation, but the narrative codes of suspense, enigma, irony, comedy and causality, to name but a few. To write a novel is to conduct imaginary personages through imaginary space and time in a way that will be simultaneously interesting, perhaps amusing, surprising yet convincing, representative or significant in a more than merely personal, private sense. You cannot do this without projecting the effect of what you write upon an imagined reader. In other words, although you cannot absolutely know or control the meanings that your novel communicates to its readers, you cannot *not* know that you are involved in an activity of communication, otherwise you will have no criteria of relevance, logic, cohesion, success and failure, in the composition of your fictional discourse. The generation of meaning unintended by the author, in the reading process, is dependent on a structure of intended meaning: the Hilary–Joy equivalence in *Small World*, for instance, is brought into play partly by the percipience of A. S. Byatt and partly by the fact that that novel is by intention full of doubles and pairs and symmetries and heavily connotative names. Perhaps I may conclude by repeating what I have written elsewhere:

> As I write, I make the same demands upon my own text as I do, in my critical capacity, on the texts of other writers. Every part of a novel, every incident, character, word even, must make an identifiable contribution to the whole... On the other hand (there is always another hand in these matters) I would not claim that, because I could explicate my own novel line by line, that is all it could mean, and I am well aware of the danger of inhibiting the interpretive freedom of the reader by a premature display of my own, as it were, 'authorised' interpretation. A novel is in one sense a game, a game that requires at least two players, a reader as well as a writer. The writer who seeks to control or dictate the responses of his reader outside the boundaries of the text itself, is comparable to a card-player who gets up periodically from his place, goes round the table to look at his opponent's hand, and advises him what cards to lay. ('Small World: an Introduction; in *Write On* 1985).

It might be profitable to pursue this idea further: the novel not as communication, not as production, but as play. But that would be another lecture.

FURTHER READING

Benjamin, Walter. *Illuminations*, London: Jonathan Cape, 1970.

Chatman, Seymour. *Story and Discourse: Narrative Structure in Fiction and Film*, Ithaca: Cornell University Press, 1978.

Clark, Katerina and Holquist, Michael. *Mikhail Bahktin*, Cambridge, Mass: Harvard University Press, 1984.
Lodge, David. *Modern Criticism and Theory*, London: Longman.
Lodge, David, ed. *Twentieth Century Literary Criticism*, London: Longman, 1972.
Young, Robert, ed. *Untying the Text: A Post-Structuralist Reader*, London: Routledge and Kegan Paul, 1981.

6

Communication without words

JONATHAN MILLER

Wittgenstein once asked what was left over after one subtracted from the sentence 'I raise my arm' the sentence 'My arm goes up'. A comparable question about communication might go something like this. 'If I were to set aside all those communications which are expressed or expressible in words – written, spoken or signed – what would be left over?'

For anyone who regards *language* as the canonical form of human communication, the answer would probably be 'Not much is left over' and the residue, such as it is, is either a redundant supplement to words – something which the telephone shows we can do without – or else a sadly impoverished alternative which we are sometimes compelled to use when circumstances make the ordinary use of words awkward or impossible.

On the other hand, for those who regard language with suspicion, especially written or printed language, on the grounds that it misleads and confuses as much as it informs and expresses, eliminating words and sentences exposes a level of communication of unsuspected richness, one in which human beings express their true meanings. The idea is that articulate language is a barrier to rather than a medium of communication and that if only this barrier could be removed, human beings would revert to a golden age of wordless, heartfelt communication. This attitude to non-verbal communication has been encouraged by the popularisation of right-brain left-brain studies and amongst those who sponsor the soft primitivism that I have just referred to it is widely assumed that the verbal capabilities of the left cerebral hemisphere have been over-developed by a culture which puts too much emphasis on linguistic finesse and that the expressive repertoire of the supposedly holistic right hemisphere has been dangerously neglected as a consequence. In fact there are those who go even further, insisting that favouring the verbal capacities of the left hemisphere not only conceals but actually deforms and disables right-sided accomplishments. The most widely publicised example of this claim is to be found in Betty Edwards's best-selling *Drawing on the Right Side of the Brain*. In this astoundingly popular and not altogether unpersuasive book, Miss Edwards sponsors a pedagogical programme designed to diminish the influence of linguistically determined ways of seeing the world. Her argument is that by learning to overlook those parts of the world which are easily nameable we can revert to a mode of perception more favourable to successful drawing. Here is one of her recommendations.

> The left hemisphere is not well equipped to deal with empty spaces. It can't name them, recognise them, match them with stored categories, or produce ready made symbols for them. In fact the left brain seems to be bored with spaces. They are therefore passed over to the right hemisphere. To the right brain, spaces and objects, the known and the unknown, the nameable and the unnameable are all the same. It's all interesting.

And so on. There are some other interesting strategies recommended in the book and as someone who has always been frustrated by his inability to draw nicely I am bound to admit that the exercises suggested by Miss Edwards have brought about an unexpected improvement in my performance as a draughtsman. Now whether this has anything to do with a conflict between right and left halves of the brain is not really the issue here. In any case I don't intend to discuss the visual arts as an example of non-verbal communication and I only introduce the topic of drawing to illustrate the extent to which antagonism to language has infiltrated itself into at least one important department of educational theory. There are other examples though. Although the advertising industry is almost promiscuous in its use of verbal slogans, the creative emphasis falls more and more upon the persuasive power of imagery – slow motion shots demonstrating the lustrous lightness of newly washed hair, or the soft resilience of freshly laundered towels – in fact it would be tiresome to list the repertoire of non-verbal devices deliberately designed to by-pass a critical vigilance based upon language.

A comparable tendency is to be found in the theatre. Inaugurated in the 1960s with the rediscovery of Artaud's manifestos in favour of the so-called Theatre of Cruelty, drama in the last quarter of the twentieth century displays a noticeable interest in bizarre expressionistic decor, extended pantomimic gestures and sometimes a cacophony of non-verbal sounds. In the increasingly popular idiom of so-called 'performance arts' actors and audiences revel in non-verbal excesses in the belief that such behaviour addresses itself directly to the human soul and that all other forms of traditional theatre are disgustingly 'literary'. This repudiation of language is often associated with the more romantic forms of political radicalism, the idea being that language is one of several devices by which the ruling elite manipulates cognitive structures to its own advantage, and that it is only by storming the Bastille of linguistic tradition that human beings have any hope of being restored to a state of primaeval egalitarian fellowship. This attitude is one of the things that has given non-verbal communication such a bad name and since it already has a somewhat shaky reputation due to the fact that it has no powerful theory associated with it, its academic credibility suffers in comparison to that of formal linguistics. In fact even if one succeeds in dissociating oneself from some of the more romantic claims that are made on its behalf it's easy to get the discouraging impression that communication without words is after all a residual topic and that once orthodox language has been subtracted all that is left is a rubbish heap of nudges, shrugs, pouts, sighs, winks and glances – or to put it another way that non-verbal communication is simply the behavioural exhaust thrown out of the rear end of an extremely high-tech linguistic machine.

And yet...*is* it all that easy to subtract language in the first place? Can one really strip away the lexical component leaving behind a non-verbal residue which has nothing to do with communication in words? The fact that one can commit words to paper without any apparent loss of intelligibility suggests that there is, in fact, a clean division between the lexical and the non-verbal component of human communication, and that the so-called kinesic variables such as facial expression, posture, and hand movements are just optional extras. But this conclusion overlooks the fundamental distinction between the meaning of an utterance and the meaning which the utterer wishes to convey by means of that utterance. Because although it could be argued that what an *utterance* means is readily recoverable by anyone who can read printed English, it is important to understand that what the speaker wishes to express is more often than not defined by the factors which get lost in the process of transcription. The problem is that writing was not developed in the first place to preserve the meanings of talk or conversation. It was developed originally to promulgate priestly or legislative initiatives, and since these were collective and in some sense impersonal productions, what the writer meant was to all intents and purposes recoverable from what he wrote down. If there was any ambiguity, that is to say implications which might escape the first or indeed many subsequent readings, they were not the ones which would have made themselves more readily apparent if some form of graphic representation had preserved the tone of voice, the facial expressions or the hand movements of their author. So that there was no incentive to develop a notation designed to represent the non-lexical parts of an utterance. In fact, the notational shortcomings of writing only became apparent when authors tried to reproduce the talk of individuals. Then, and perhaps only then, the difficulty of identifying speech acts becomes apparent.

The notion of speech acts was introduced by the Oxford philosopher J. L. Austin, who pointed out that in uttering this or that well-formed sentence a speaker is doing something over and above expressing its literal meaning. He or she may be stating, describing, warning, commanding, apologising, requesting or beseeching. In fact, according to Austin there are more than a thousand of these acts which are performable in English, and unless the hearer or reader recognises which of these is being expressed by the utterances in question he or she has missed the point.

Of course, the identity of a speech act, its illocutionary force as Austin calls it, is often made apparent by an explicit lexical indicator. 'I *warn* you that I will take steps to prevent you' or 'I *promise* that I will be there on time'. And in such cases the non-lexical cues – finger waggings, handshakes and so forth – are indeed superfluous, and the printed text preserves everything that the utterer intended to convey. But for each of the thousand or so explicitly identifiable speech acts there are just as many for which there is neither a name nor a lexical indicator. And in that case the only way of identifying them with any accuracy is to hear them spoken and to witness the non-verbal behaviour with which they are preceded, accompanied or followed. A playwright will often do his best to supply this non-lexical information by telling the reader that the character shrugs, winks, or looks heavenwards as this or that phrase is uttered. A novelist can be even more helpful by saying that the phrase in question was spoken waggishly, or grimly, or that it was snapped out as the character turned angrily on his heel. However, the grain of this behavioural notation is unbelievably coarse and one is often surprised by the extent to which two performances of the same written utterance can differ – even when the actors in question are apparently following the same instruction with respect to intonation, facial expression or manual gesture. The result is that instead of trying to recover the often indeterminable illocutionary force intended by the author for this or that character, the actor finds himself inventing someone who might have wished to express this or that speech act by means of the speeches assigned to him in the text. In which case the non-verbal concomitants of the various utterances are improvised as if for the first time, and in the best of all possible productions an unforeseeable Lear, Macbeth or Rosalind emerges in performance, and the speeches come across expressing meanings which would have been hard to foresee from reading the bare text.

The point I am labouring at such length is that there is a large and complicated repertoire of non-verbal behaviour without which it is impossible to communicate meanings through the medium of spoken words and although it is tempting to regard this non-lexical repertoire as something which can be painlessly removed without any significant loss of meaning, the experience of reconstructing talk from a medium in which the representation of this aspect of speech is so poor is a salutary reminder of its importance.

Up to this point I have concentrated on the way in which non-verbal behaviour helps us, as Austin would say, 'to *do* things with words'. I would now like to turn my attention to something which is in a sense a mirror image of what we have been considering. How can we 'say things with deeds'?

There is, of course, a sense in which all our actions or deeds speak louder than words, and that everything we do – or fail to do, for that matter – is open to interpretation and therefore counts as a communication. In fact, it doesn't have to be anything properly identifiable as a deed to communicate interpretable evidence. A blush, a hangdog posture or a limp handshape can all convey information and experts in so-called body language – horrid phrase – have compiled long lists of postures and gestures from which an observant onlooker can glean some information about the attitudes or intentions of others. Mere 'presence' can speak volumes. Someone who turns up at an occasion which is known to be an ordeal for him communicates information whether he wishes to or not. His unexpected presence may be interpreted, rightly or wrongly, as a deed deliberately intended to express his courage or defiance. A well-known alcoholic who unexpectedly turns up at a cocktail party may inadvertently communicate the fact that his sessions with AA have given him newfound confidence in his self-control. But his turning up at such an occasion may be an explicit act of communication – a way of saying without words that he can now resist the blandishments of the bar and that his friends and colleagues are to regard him as a reformed character. It is important to distinguish, as far as one can, between behaviour which wordlessly *betrays* information, behaviour, that is, from which an onlooker may *glean* something, and non-verbal behaviour which is performed with the express purpose of *communicating* this or that information. Here is another example. Someone who manages to read in a noisy, crowded room may inadvertently communicate evidence as to his enviable powers of concentration. But since the act of reading monopolises his attention, he is by definition 'dead to the world' and therefore unaware of that fact which his behaviour communicates. In contrast I have chosen the following passage from *Barnaby Rudge*:

> 'How do you find yourself now, my dear wife?' said the locksmith, taking a chair near his wife (who had resumed her book), and rubbing his knees hard as he made the inquiry.
>
> 'You're very anxious to know, an't you?' returned Mrs Varden, with her eyes on the print. 'You, that have not been near me all day, and wouldn't have been if I was dying!'
>
> 'My dear Martha –' said Gabriel.
>
> Mrs Varden turned over to the next page; then went back again to the bottom line over leaf to be quite sure of the last words; and then went on reading with an appearance of the deepest interest and study.
>
> 'My dear Martha,' said the locksmith, 'how can you say such things, when you know you don't mean them? If you were dying! Why, if there was anything serious the matter with you, Martha, shouldn't I be in constant attendance upon you?'
>
> 'Yes!' cried Mrs Varden, bursting into tears, 'yes, you would. I don't doubt it Varden. Certainly you would. That's as much as to tell me that you would be hovering round me like a vulture, waiting till the breath was out of my body, that you might go and marry someone else.'

Unlike a person whose *actual* reading betrays his powers of concentration, Mrs Varden's *pretended* reading prevents her from actually reading, because in order to monitor and enjoy its communicative effect it would be impossible for Mrs Varden to accomplish the deed of reading in earnest. But not all pretended deeds have to fall short of their normal function in order to accomplish their communicative purpose. Take the example of the burglar in Austin's famous essay on pretending – surely a classic example of saying something with deeds as opposed to doing something with words. A burglar is inspecting a window with a view to breaking and entering, but in order to make his interest look innocent he pretends to be cleaning the windows. As it happens, the most convincing way of pretending to clean a window is to actually do so.

The most observant reporter of saying things by means of deeds was the late Erving Goffmann, and it is to his work that I would like to dedicate the rest of this lecture. I do so as a grateful tribute to someone who has liberated the study of non-verbal communication from the dead hand of ethological reductionism.

A central feature of Goffmann's approach to non-verbal behaviour is his assumption that it is to be visualised against the background of institutional norms which create the salient facts of social life. Without an appreciation of these norms it is almost impossible to make sense, let alone describe, much of the conduct which characterises our mutual involvements. According to Goffmann, what lends credibility to our concepts of personal self is the recognition of certain rules or conventions which limit the claims we can expect to be acknowledged with respect to freedom from untoward threat, interference and so forth. We venture into public life protected not so much by the sanctions of formal law but by an unwritten charter of civil rights which assigns us both access *to* and independence *from* others with whom we come into contact.

Those who lay claim to these rights and expect to have violations recognised and remedied, know that they undertake reciprocal obligations and will be expected to provide appropriate remedies if they are guilty of infraction, even if innocently. In our transit across public places we rely on others recognising the rules which assign us the right to proceed without being inconvenienced by impudent stares or unsolicited conversational openings. On the other hand we also proceed on the assumption that we have some measure of personal access to others if the occasion unexpectedly requires it and vice versa. And that if such openings ensue that there are supportive rituals which allow us to engage in them without offence and terminate them without insult. In return for such a privilege we implicitly acknowledge that there are reciprocal obligations incumbent upon us.

What this means is that the individual in public feels obliged to broadcast an unceasing stream of non-verbal signs, intended to inform others, whether they be acquaintances or more often otherwise, of the place which he or she expects to have in the undertakings which follow. By means of such conduct, we inform one another about the legitimacy of our presence, the innocence of our motives and our readiness to grant access or co-operation if the situation arises. And at times or places where our actions are likely to be misinterpreted the intensity of this indicative behaviour increases.

As Goffmann points out these signs have been neglected or disparaged as trivial items. Or even worse they may be misdescribed as vestigial bequests from our primate ancestry – yet another example of the naive reductionism which sometimes passes as orthodox science. Where Goffmann scores is by allocating scientific importance to the *moral* representation of self in everyday life.

Consider for a moment the question of legitimate presence. In places, where anything short of purposeful warning might be misinterpreted as either suspicious loitering or aberrant vacancy, the normal person feels obliged to put on a show, which tells anyone who might be watching that orderly motives are in hand. He or she will glance ostentatiously at his watch, as if to indicate that an expected arrival is late for an appointment and if he happens to meet the glance of a passer-by, he will more often than not look once again at his watch and cast a long-suffering glance at heaven; as if by recruiting sympathy for a *familiar* predicament he will pre-empt any suspicion of more suspect motives. Such behaviour will perhaps be even more pronounced if the innocent loiterer happens to have stationed himself at places where his presence might be misinterpreted. Washrooms and lavatories are classical locations for such conduct. In Men's rooms, which are the only ones from which I can report personal experiences, there are elaborate rituals for avoiding the impression of suspect motives. A concentrated stare at the white tile immediately ahead of one usually takes place when someone unknown unexpectedly occupies the stall alongside – sometimes accompanied by the onset of a tuneless and preoccupied whistle – anything to avoid creating the impression that one might be showing an untoward interest in the UG equipment of one's neighbour. Of course such behaviour won't wash, and I use the word advisedly, if the neighbour happens to be a colleague. For in *that* case, the elaborate precautions to avoid eye contact could be read as a suspicion of *his* motives and thereby create a second order of virtual offence.

For obvious reasons, the situation is less fraught with the risk, in the purposeful *va* and *vient* of open corridors. Nevertheless even here an unremitting etiquette prevails. It is an etiquette in which the participants tacitly assume that there are reciprocal obligations with respect to right of way, freedom from inquisitive glances and capricious encroachments upon privacy. At the same the management of eye movements leaves room for the possibility of accesses which can and often do develop into what Goffmann describes as focussed encounters; episodes which are themselves introduced and terminated by rituals of greeting and farewell. Such episodes may, of course, be confused as passing acknowledgements, but the readiness to exchange such signals is *one* of the ways in which we register the normality of the passing scene and it is when we encounter consistent anomalies in the broadcast that we begin to suspect and perhaps report something odd.

In institutions such as hospitals or the BBC, where colleagues and acquaintances run the risk of passing one another many times in the same morning, the ritual resources for handling brief encounters are often over-stretched. First and second encounters can be managed by conventional openness – a third meeting may necessitate a humorously resigned grin – a fourth can be handled by pretending to be wrapped in thought – a fifth may require some dramatised horseplay such as play-acting a Western duel. And you've all seen and probably participated in the scene where a sequence of such meeting is brought to its climax by one partner coming right out with the movie cliché 'We can't go on meeting like this' or less effectively 'Long time no see!'

All this, as Goffmann points out, presupposes three levels of normal functioning:

> (a) The recognition of the fact that an individual is a potential source of alarm, inconvenience, offence and encroachment.
> (b) Recognition of the fact that each individual has both the obligation to minimise these aspects of himself AND the capacity to do so.
> (c) Recognition of the need to perform remedial work if one recognisably infringes any of the norms which one intuitively regards as binding.

The point is that almost any configuration of events with which an individual is likely to be associated in public carries the risk of a *worst* possible meaning which might reflect unfavourably upon him, and it is a sign of intact mental functioning that one recognises this risk, without of course being incapacitated by the thought, and at the same time that one is equipped to perform repair work if and when infractions occur.

It is, I think, in the analysis of this so-called remedial work that Goffmann is at his most imaginative and productive. One of the things that makes his account so useful – so much more than the anecdotal triviality of which he is so carelessly accused – is his ability to compare and contrast this *informal* repair work with the formal structures of explicit legal process. As in law there is an orderly sequence of offence, arrest, remedy and reconciliation. But what distinguishes these *interchanges* is the fact that the offender is so often the first to recognise that an infraction has occurred and usually initiates the appropriate repair work without being asked to do so.

An even more important distinction is the fact that the remedial work is expressive rather than productive. In other words the remedial performance is designed to restore a favourable image of the offender as opposed to offering substantial compensation to the offended.

Taking his cue from yet another of Austin's philosophical essays, the famous and often reprinted 'A plea for excuses', Goffmann distinguishes various forms of remedial ritual, of which the first is the so-called *account*. In this, the offender re-describes his or her act so that its offensiveness may be overlooked or discounted. It may take the form of an explicit explanation. Someone, for example, who finds himself in the embarrassing situation of seeming to have winked at an unknown passer-by may offer the account that he has some grit in his eye – this often accompanied by a flurry of overacted eyelid-rubbing and nose-blowing. In this way he re-establishes his image as an altogether innocent victim. Of course, one has to be careful in this context to recognise that many of the infractions I'm referring to are not necessarily offences AGAINST others – but represent errors of performance, imperfections which reflect badly on the offender – so that one undertakes remedial work, NOT for the purpose of making amends but to re-draw the picture of oneself so that it corresponds more closely to the one which one would *like* to project to the world at large. So important is this consideration, and it would be perfunctory to regard it as *mere* vanity, that it may motivate performances to anonymous and usually unconcerned strangers. You only have to think of the otherwise incomprehensible behaviour of someone who hails a cab with a flailing gesture of the outstretched arm and who, having failed, then feels it necessary to provide an *account* of what happened by using the same hand to smooth down the hair. Or in Goffmann's own example of the man who trips in the street, to his own and no one else's inconvenience, who then feels it necessary to retrace his steps and conscientiously examine the sidewalk – as if to establish the impression that the fault lies in the pavement and not, as might otherwise be suspected, in the nervous system of the person concerned. The point is, that whether it's addressed personally, or all round to anyone who might be watching, whether it's verbal or mimetic, the function of an *account* is to correct a potentially unfavourable impression of oneself which an infraction of the unwritten rules might produce.

And the same principle applies to *apology*, although as Austin pointed out in his essay, the *logic* of apology is not the same as that of accounts. In making an apology one accepts blame for what has happened, but at the same time one tries to convince the injured party if there happens to be one, or the world at large if not, that the error is not to be taken as representative of the *real* self.

Apology, in other words, is aimed at convincing anyone interested that the miscreant recognises his fault, and by *that* token alone, is to be regarded as someone whose *typical* tendency is to observe the conventions. Such a performance may be verbal or non-verbal. In circumstances when words are inappropriate or impractical, the apology may take the form of an elaborate pantomime of contrition. On entering a small seminar room, where a meeting is already underway, the show may take the following complex form. A self-uglifying expression of humility – plus an elaborate show of stealthiness which is as good as saying 'Yes, I am late – and please pay attention to my performance of humbly *not* wishing to be paid attention to', i.e. Here's me entering as unostentatiously as I know how – so you can see how much I regret my rudeness!'

A comparable version of this is the face made by someone who barges into a room unannounced expecting to speak to a friend, only to find that this friend is engaged in an intimate professional consultation with another colleague. Although a verbal apology would probably fit the bill, the offender may feel constrained to act the fool he expects to be accused of being. Hence an otherwise unintelligible grimace. Or the actor who stumbles over his words for the *second* time at a rehearsal. He will often apologise by overplaying the spastic idiot everyone around must suspect him of being.

There are also, I think, concealed apologies included in the otherwise straightforward rituals of farewell. As Goffmann points out, the *end* of conversational encounters carry an increased risk of creating offence – in the sense that careless or perfunctory termination may convey the misleading impression that one couldn't wait for the session to end and that as far as *one* was concerned the whole episode was a waste of time. On occasions where this is felt to be a risk, preventive apologies may be issued in the form of prolonged negotiations to meet again soon. Or anything to avoid the potentially offensive gesture of actually *leaving*!

This of course raises the question of the apologies and/or accounts which accompany failed farewells. The situation I'm thinking of is this. One's been talking with a small group of colleagues. Because of an appointment or whatever one has to leave before the group as a whole breaks up. Having successfully manoeuvred an inoffensive farewell, one discovers that one has left a book in the room. Now try and visualise the risks of re-entry. First, the offence to onself. This is usually surmounted by merely explaining 'Left my book'. But since one may suspect that one'll be thought a fool for having done so, it may be necessary to overact being a fool and murmur 'Forget my own head next'. Perhaps this show is reinforced by miming a stumble or a mindless struggle with the door on leaving yet again. But the situation is complicated by the knowledge that in one's all too brief absence the space left by one's departure is already in process of closing over – new topics are in hand and one might create offence to the members of the reconstituted group by seeming to re-insert oneself. Once again the tip-toe manoeuvre – but this time it's not quite an apology so much as an unsuccessful account. An account which tries to convey the impression that you're not there at all. And so forth.

Now...in my enthusiasm for *anecdotal* aspect of all this I have neglected to mention the other half of Goffmann's analysis of remedial procedure. I am referring of course to the process of *closure* – that is to say the ritualised responses, whereby the injured party acknowledges and accepts the accounts or apologies, thus allowing social activity to resume its productive course. If this so-called round is left incomplete the offender, virtual or actual, is left hanging in the air, uncertain as to his moral status in the undertakings that will follow. These replies may seem too trivial to mention – a nod, a murmur, 'that's OK' or whatever, but if these signs are not provided the offender is left with the uneasy sense that his or her offence, trivial or not, is permanently entered in the criminal record.

It is, I think, one of Erving Goffmann's most lasting achievements to have made these interchanges both visible and intelligible. And what makes his analysis so attractive is the fact that he has resolutely turned his back on the temptations to reduce what he has seen to some supposedly more fundamental principle of animal behaviour. As far as he was concerned, what we are witnessing in these exchanges is the expression of the distinctly *moral* part of human nature. In his own words,

> If we examine what it is one participant is ready to see that other participants might read into a situation and what it is that will cause him to provide ritual remedies, followed by relief for these efforts, we find ourselves looking at the central moral traditions of Western culture.

FURTHER READING

Goffmann, Erving. *Relations in Public: Microstudies of the Public Order*, New York: Harper & Row, 1971.

7

Music as communication

ALEXANDER GOEHR

TOPOGRAPHY AND POLITICS

As soon as we talk about music as communication, we imply a topography, and arising from it a politics. Imagine musical progress as space: the topography is determined by the way the various participants associated with it relate to each other: at one edge the original sender of the music, he who made it or composed it, at the other its recipient, the listener; and between these two are the musicians, the singers and players who manufacture sound, and nowadays even the recording engineers who effect its electronic transmission. Arising from this, the politics is played out between the sometimes complementary and sometimes conflicting concerns of these three (or four) groups. Each has its own values: which is to predominate? Should the entire musical process be regarded as the recreation of the composer's original intention? Or should we bear in mind that he is quite a recent phenomenon; that music existed for centuries without any observable separation between composer and performer. Or then again, is music really a performing art? Compositions, if they have a separate identity, can be regarded as of no greater importance than a set of instructions: to be followed, but adapted to circumstances. Fidelity to the composer's intention is then less important than the performer's own feeling expressed in the way he plays his instrument. Or again, should not everything be evaluated from the point of view of the recipient? After all it is he who pays! Never mind who makes it or how it's played: let us ignore how the music came to be, let us consider it to be no more and no less than what it is heard to be.

Figure 1 Music as communication: Oscar Peterson and company (Photo by Dennis Stock: Magnum Photos Inc, New York)

The relative weight to be attached to these considerations does not remain constant at all times. No one hierarchy can be assumed as valid for the whole history of music.

This is because music serves different functions at different times and in different places. Although tempting to consider it as no more than a system of perceptible sounding structures, an understanding of the psychological process does not of itself explain why people actually cultivate music, nor what they see in it. Nor does it account for the fact that some kinds of music are simply not understood and will be rejected out of hand, and not merely because they are judged to be aesthetically poor. In fact as much confusion is generated by regarding music as a system of sounding structures, as by the sentimental old notion that music is a universal language, which inevitably presupposes that one's own musical language is universal, because were it not, music could not be considered in this way at all. The physical sounds made by blowing, singing, plucking and bowing and banging are more or less the same wherever they occur; there are after all only a limited number of ways of creating them. But everything that might be considered to make up the way music is understood (could I even say, loved): the being moved by it, the being aroused or the admiring of skill, depends on specific ways of ordering such sounds into specific and familiar contexts. As with food, the raw materials are the same everywhere; but this does not make us like everybody's cooking, even if it's as nutritious as our own.

Music is as bound to context as is language. Whereas we don't take seriously the person who, without a knowledge of a particular language, supposes that he can extrapolate meaning from phonemes, we do tend to assume that music, of whose rules and practices we know little or nothing, will somehow succeed in evoking images of a purely sensual and vaguely associative kind. But where the context is established (and we are part of it) and where social function is predetermined, real communion can be taken for granted.* Here and now, in our society, different kinds of music emanating from different periods of history and different contexts all coexist; and here communication has to be effected largely without reference to the original function of the music. Without knowing whether a piece of music is intended to celebrate a funeral or a marriage, we nevertheless suppose that we are able to understand it. Inevitably this unleashes a process of continual rejigging and adapting of such different kinds of music to fit into the way in which music does function and is understood by us. For we should not delude ourselves into supposing that, even in what is clearly a complex society and one in which simple communal values have been submerged, there would not now be unspoken conventions and customs which determine functions. Inevitably the fact that music is extracted from its original context to make it fit into our own distorts it and changes its identity.

FUNCTION AND AFFECT/LACK OF FUNCTION: FAILURE OF AFFECT

Wer Gefühle als Wirkungen der Musik davontragt, hat an ihren gleichsam ein symbolisches Zwischenreich, das ihm ein Vorgeschmack von der Musik geben kann, doch ihn zugleich aus ihrem innersten Heiligtum ausschliesst!

Nietzsche, 'Über Musik und Wort' (*Werke*, Insel, vol. I)

(Whoever derives feelings from music as its effects has in them, as it

* I am indebted to Sir John Lyons for drawing my attention to Sir Basil Maine's differentiation between 'communion' and 'communication', reflecting as it does the difference between community and society.

were, an intermediate realm of symbol, which while providing a foretaste of music, at the same time denies entry to its innermost sanctuary!)

Comparison between music and language (or the belief that music is some kind of language) always falls on the issue of semantics. Is music a set of syntactic structures which constitutes a vehicle for transmitting affects or meanings? Or does music (in what was originally Stravinsky's pithy formulation) 'express nothing but itself': that is, does not communicate things outside itself? The argument hinges on whether or not intervals between pitch-levels, rhythms, melodies, modes and harmonies of themselves convey feelings or evoke images of place or time. Even if it is popularly believed that they do, it does not thereby follow that the capacity to do so is part of their nature. If there is a convention, established by whatever means, that an interval conveys a feeling, that feeling will be automatically aroused by the sounding of that interval. Such conventions are part and parcel of the very functions which the music originally served. But when the music outlives these original functions its conventions are only half (if at all) remembered, and only a vague penumbra of affect and emotional meaning survives.

But the discussion of the origins of musical affect, whether they are to be found in nature or convention, is an old chestnut, and I doubt whether there is much to be gained from turning it over. Most performers and listeners still do consider music in terms of its affects, even if academic attention is principally focussed upon problems of structure. Crude as it may sound, it hardly matters whether a piece expresses joy or sadness, is religious or witty because the intervals, harmony or rhythm genuinely evoke such emotions, or whether the convention of these emotions are merely a convenient tag for describing the effect of music in words. Insofar as such characterisations of musical affect do actually survive and are judged useful I am happy to leave it at that.

But how explicit and precise can such individual musical affects be? A musician aspires to total exactitude; to play a wrong note, or the right note at the wrong moment, or to play loudly what is marked quietly, either renders the passage grammatically incorrect, or changes its identity. But if such a discipline were deemed to hold true for the listener, most of what is normally assumed to be an understanding of music would be thrown into question. Few are those who, with hands on heart, can claim to absorb each and every musical event with equal and undivided attention. In reality, music transmits different grades of information, and each requires a differing degree of attention. Intervals and motifs need to be precisely registered,

Figure 2 Music as communication? The Music Lesson by Marcel Frischmann

whereas long-term continuity is often perceived in a dreamy state. If one asks for a street direction, there is no point in listening to only a small part of the answer. But aesthetic pleasure and consequently some meaning can be obtained from the semi-attentive reading of a complex poem. There what is actually called 'its music' is communicated, even when many of its specific allusions are missed. This suggests that music is sometimes more appropriately perceived *grosso modo*, from its passing flow towards its detail, rather than from its detail toward its totality. Paradoxically, as is the case with much of the music of this century, the more differentiated and the more complex the structure of a piece, the more the tendency to be satisfied with a lesser amount of precise information. Though music teachers train the aural capacities and memories of their pupils (and test their work out of an imagainary 100 per cent), they well know that in the real world, listeners' degrees of attention vary considerably from moment to moment, and that this reflects not only their powers of recognition and differentiation, but the span of their concentration and the amount of information which they absorb at any particular moment. Whether or not an individual listener has absolute pitch or very good relative pitch, and whether he (or she) is capable of perceiving fine harmonic and rhythmic differentiations, may determine the quality of listening, but not necessarily the amount or quality of either understanding or aesthetic pleasure obtained. Were it to be assumed that the ideal listener is the one who perceives each and every sound, not only would there be few of him, but it would seem to imply that music all over the world had always and did now serve an identical function; that it was to be listened to with total attention as if it conveyed a precise communication. Or it would resemble an abstract activity rather like playing chess. But, in reality, in music, as indeed more generally, the dimension of meaning is shaped by role. It makes no more sense to claim that a specific unit of meaning can be extrapolated from one event in a musical composition, than to suppose that different pieces serving the identical social role – every march, hymn or love song – are sufficiently defined by that role and therefore convey an identical set of meanings to each and every listener. Arguments about affect and function have to be modestly conducted in terms of more or less, both quantitatively and qualitatively.

Despite everything that can be advanced in its defence, an 'argument in favour of imprecise hearing' (as one could term it) seems a bit complacent and even condescending. If a composition is in actual fact highly differentiated in its detail, and consequently charged with great intensity, we not only owe it, but are likely to give it, our full attention. Its very effect is determined by the degree of differentiation of its detail and, if this is not to be precisely observed, of what else can the communication consist? How is the music to make its proper effect without it? Proper, I say; and 'proper' suggests that each or every isolated musical event is intended to have a specific effect, excluding all others, and that every sequence of such events has an equally specific effect; and therefore a whole movement and a whole work has a particular overall effect; and that there can be only one such. If a piece is not to be wholly perceived as the sum of its parts, it will be a partial failure on the part of either composer or listener, or both. Such failure leaves the listener with a sense of his own inadequacy. If this is so, dreamy semi-attentive listening for the pleasure it gives might now become associated with mediocrity and guilt. It is cheap and easy, the line of least resistance, and to be scorned as a feckless and ultimately frustrating hedonism and, as such, a debasement of proper artistic purpose.

> Pendant que des mortels la multitude vile,
> Sous le fouet du plaisir, ce bourreau sans merci,

> Va cueillir des remords dans la fête servile...
> (Baudelaire, *Recueillement*)
> Now while the common multitude strips bare,
> Feels pleasure's rat o'nine tails on its back
> And fights off anguish at the great bazaar...
> (translation: Robert Lowell, *Imitations*)

Or Karl Kraus, even more directly and aggressively: 'I detest the ugliness of a pleasure-savouring crowd who, after a suffocating day at the office, raises the lowered blinds over its soul to let in the air of culture.'

Such highminded expectations for art, and aggressive reactions when they seem to be frustrated, lead the artist to despair of his audience, to seek out the commitment of the like-minded and finally, when all else fails, leaves him with his own imaginary double, his mirror image, the ideal listener made in the likeness of himself – art can exist without the need to communicate anything at all to another and becomes entirely solipsistic.

MUSIC WITHOUT LISTENERS

Even without an exclusive retreat into the self, it is in fact viable to consider music as having a justifiable existence without listeners. To this day in many communities music forms part of a ritual in which all present participate, and no clear differentiation exists between performer and listener. Here then 'communion' would most appropriately describe the way the participants relate to each other, each and every individual performing his allotted role. The gap in the original topography between sender and recipient is dissolved and no longer exists. But quite as much as with music specifically intended to communicate, the individual participant here too is provided with a range of different psycho-physical sensations.

Let me give a simple example: many years ago at a private party – in fact it was after the first performance of Olivier Messiaen's *Turangalila-Sinfonie* in London, and the first time I met my future teacher – we played a clapping game. The idea, Messiaen explained, was that each person should clap in turn, contributing to what all together should come out as a regular and smooth sequence of claps. Of course this game could be endlessly complicated, but the point is that the participant not only perceives the chain of claps as a resultant, but feels the rise and fall of tension as he prepares to make his clap, matching it in sound to his neighbour's, then placing it correctly and, as it were, leading to his successor's. To get it right, each participant not only has to 'know his part' but has inwardly to sing the implied resultant. This periodic repetition is the simplest form of the game, and is at one and the same time the mother of each individual clap and the final form of all the claps put together. It makes the world of difference whether one is to be a participant or an onlooker of the game. Though the identity of the claps is the same in either case, the feeling they convey is completely different. This has an ethical dimension too: as when Plato observes the merits of musical activity, putting in Protagoras's mouth the notion that by 'making harmonies and rhythms quite familiar to the children's souls, they may learn to be more gentle and harmonious and rhythmical and so more fitted for speech and action; for the life of many has need of harmony and rhythm in every part'. Appropriately (for this discussion) Aristotle raises the question as to whether children are to learn by singing and playing themselves, or merely from listening to others. Franz Kafka writes in his *Notebooks*: 'The law of the Quadrille is clear, all the dancers know it and it is valid for all times. But one or other of the hazards of life, which ought not to, but over and over does, occur, brings you alone out of step. But you do not know it, you know only your own bad luck.' Here, he finely expresses what I have merely called a rise or fall in tension, equanimity and harmoniousness when an action has been correct, but indeed 'bad luck' when it goes wrong.

I felt my own 'bad luck' when, many years after the party where I clapped with Messiaen, this time at a gathering of amateur Gagaku players, in a temple in Tokyo one evening, I was assigned to play the Taiko (bass drum). I was told to repeat two strokes of the drum regularly: one masculine and one feminine (strong and weak). All I had to do was play one harder, more emphatically, than the other; but where to play them? I didn't know the composition, nor the idiom, and my strokes fell here, there and everywhere. Everyone laughed and my teacher tried to correct me. In my embarrassment however, I knew I was out of the dance and came to understand something which would never have come across to me had I been a mere listener: and that is that music is understood in a particular and authentic manner when one plays it oneself (even wrongly). What can any of us possibly make of, say, the complex structures of Congolese Pygmy polyphony? Even if we were accurately registering each and every event of the music, could we ever hope to feel it, as its performers feel it?

There is almost no point in even hoping to grasp what such music specifically conveys, assuming that the listener flies in and watches and listens, or most likely stays at home and casually listens to a recording. Any impression it leaves must be based on coincidental association or preconceptions formed far away from the spiritual environment in which the music was actually played. At worst it will seem to be nothing but a deranged jumble, at best the equivalent of a picture postcard. To get anything authentic out of it, the listener must go and apprentice himself to African musicians who perform it in the full knowledge of its context, and so stop being a mere listener. But this is impractical.

'The negro', comments Franz Kafka again, rather ruefully (again in his *Notebooks*), 'who is returned to his home from the World Exhibition, crazy with homesickness, with the most earnest mien performs the gambits which delighted the European public as the customs and usages of Africa as a tradition and a duty.'

Wherever music functions within ritual, its meaning and association is self-evident. The basic tune (and the words associated with it) are familiar, so that individual and spontaneous variants and fantasies occurring in performance can be observed and appreciated without difficulty. Within limits, complex modifications of a familiar structure may be effected in the interest of variety and sophistication and appreciated without any of the attendant problems which would arise for us who stand apart. We cannot even guess at the network of tensions and relaxations brought about by such concerted variations. Hearing and re-hearing might enable us to measure and categorise, and ultimately even form a general view about the practice of such music. But such understanding will be weightless and in no way equivalent to the sensations to be derived from being part of the music-making process.

Much the same in fact holds true for music made nearer home. Even where ritual-associated oral tradition gives way to a culture transmitted by written notation, where the taken-for-granted is replaced by an elaborate score, every performer participates in a communion, within which the whole is perceived through a network of individual tensions and subsequent relaxations. As in the clapping game, they are brought about by individual acts performed against an understood metric background. The very textbook terms used in the training of musicians: consonance, dissonance, suspension, syncopation, ornament, counterpoint, imitation, cadence, etc., all express temporal and identity relationships, thereby evidencing a pursuit of harmony, not only in its purely musical, but in its interpersonal sense. Such a concerted sequence of tensions and relaxations is actually what music is.

Meaning conveyed to the participant in terms of tension and relaxation can be further modified by the technical difficulty of that which he is required to perform. A clap is easy: but if a pitch level is difficult to produce, for example if it is in the highest register of the oboe, it will convey something other than the same pitch level which may be obtained on a violin without special difficulty. The different effect results from physical effort and concentration, not merely from the tone colour of the pitch level produced on the different instruments. Analogously a cross-rhythm (where the strong beat is articulated on the metric weak beat) conveys a special effect which disappears when it is played on the metric strong beat, though the succession of its durations remains the same (Figure 3). The effect derives from the mental effort

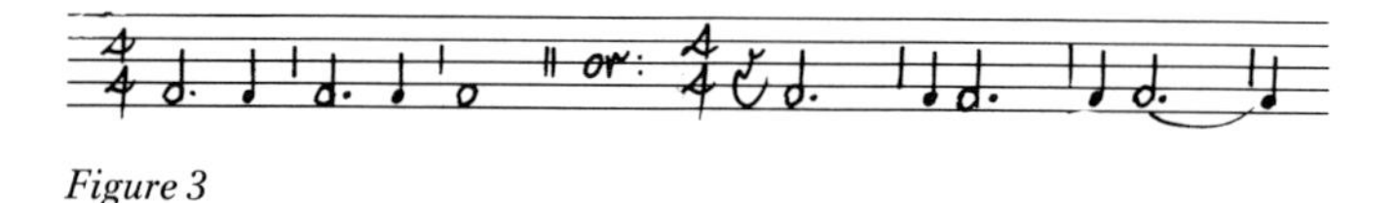

Figure 3

required to perform the rhythm against the beat. Again, quick passages are difficult on one instrument and straightforward on another, are easy in some keys and hard in others. For example on a piano: long leaps of pitch across the natural physical divisions of instruments or voice alter the effect of the sounds produced. It is difficult to hit the right note on a piano, when this requires a rapid change of arm and hand position, and sometimes it would seem logical to minimise risk of error in performance, and rearrange the hand positions to make performance safer. But to do this, is not only to cheat in a game, but subtly to alter the meaning conveyed by the gesture. Here are three *loci classici*, where difficulty conveys a special effect:

First, at the opening of Beethoven's last piano sonata, Op. 111 the pianist is required to perform a long leap in octaves with his left hand. He could use both hands, thereby simplifying the execution. He will know what he did, but do we, if, for example, we listen on a recording where we cannot see? (Figure 4)

Figure 4 Beethoven, Sonata, Op. 111

At the beginning of *The Rite of Spring*, the bassoonist is required to play in his highest register. When Stravinsky composed this, it was a dangerous and difficult passage. Today any bassoonist can play it. 'It should sound like a strangled chicken', said Messiaen. To convey that effect today, the passage would have to be transposed up a third, or even higher, on the instrument to restore the effect of the original difficulty which has now gone. This can be tested by listening to the original recordings of the passage and comparing them with current versions (Figure 5).

Figure 5 Stravinsky, *The Rite of Spring* (Boosey & Hawkes)

Finally, in the second movement of Webern's Piano Variations, the pianist is required to play notes the length of the piano apart with one single hand moving across the keyboard, then with the other – at great speed. It's awkward and risky, but quite easy when divided between both hands. The pianist knows what he is playing, but are we sure that we do? (Figure 6)

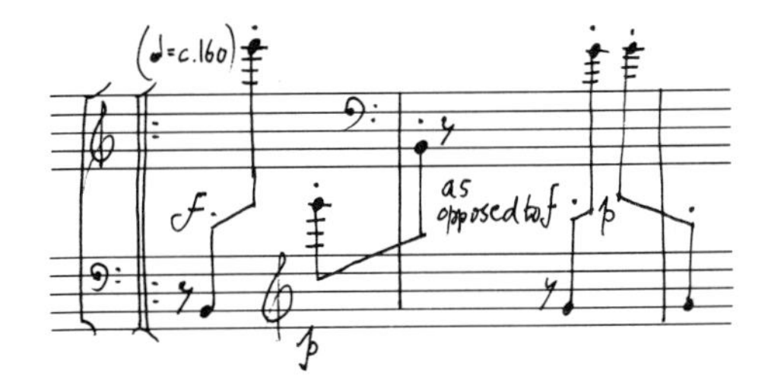

Figure 6 Webern, *Variations*, Op. 27, 2nd movement (Universal Ed.)

It may well be that such examples contribute only peripherally to the effect the music makes on the listener, and on a recording it is unlikely that such aspects could ever be fully appreciated. In a live concert the player's physical effort may be observed and might colour a perception of the actual sounds heard, in the way a hand gesture might qualify a meaning conveyed by the spoken word. The very fact that little of this physicality conveys itself to the listener must impoverish the effect of the music, making it more likely that extra-musical or symbolic meanings will be invoked as ways of sorting out the flood of information transmitted by a piece. But there is little need of such if adequate meanings can be extrapolated from the psychological effects of physical effort in actual performance.

COMMUNICATING WITH THE LISTENER

Even if much of the psycho-physical content of music inevitably remains hidden from the listener, he being presented with it, as it were, from the front of house, it would be pessimistic to think that nothing of it gets across. Possibly its tensions and relaxations, sense of strain, effort and risk inevitably do communicate without necessarily being separately distinguished. Even without it, the bystanding listener can become an active, if minor, participant in the music-making process. Like the performer, he perceives the background metre of the music, expressing such awareness by beating time (either inwardly or, more irritatingly, outwardly) and singing along. If, for example, at the opening of Mozart's G minor Symphony he is able to relate what he hears to its implied background, he is doing exactly what the performer does and thus establishes a kind of bridge between himself and the performer, feeling the tensions and relaxations of each momentary event as it moves away from and coincides again with the beats of the metre (Figure 7). This provides a physical perception of the music exactly mirroring that

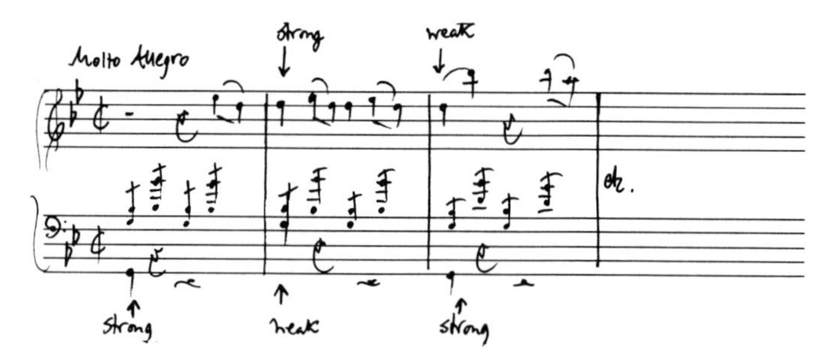

Figure 7 Mozart, G Minor Symphony, K. 550, 1st movement

of the performer. The same procedure can be applied to the perception of melody, mode and tonality: if the listener is aware of the common elements or mode of a song, he will perceive a melody in its unfolding in relation to its background. To be able to do this requires skill and experience (sometimes simply called musicality) and this constitutes an entry ticket into the musical community.

The listener might thus regard himself as a legitimate extension of the performer; and in this way abolish the gap which separates him from the sender of the music. This inevitably happens in popular music, which is monodic in form and easy to follow and memorise. But although Western art music is an outgrowth of such monodic structure, it has become so complex, that only the gifted can be relied upon to perceive background and remain conscious of it while listening. Most of us are content to form impressions from more partial and transient perceptions.

Perhaps, for this reason, people like to hear the same piece rehearsed over and over again. Even if one thinks one knows the 'Eroica' very well and has studied it carefully in all its details, somehow its full complexity inevitably escapes one at any single hearing. Possibly what is called a 'classic' is in fact a composition of such multi-dimensionality that it can never be fully perceived at a single go. Inevitably one wants to hear it again. (But this cannot be the only explanation, because the public also wants to re-hear popular songs of very little complexity, until they come out of their ears!)

The effect of Western polyphonic music of the Classic and Romantic periods can be traced back to two specific inventions. The first is the distribution of information, formerly transmitted by a single voice, between several individual voices, each in turn coming to the fore as the carrier of the main melody, and sometimes retreating to the background and contributing accompanimental or complementary material. This represents a considerable development of the simpler kind of polyphony, in which the information to be conveyed by each voice is established at the outset and remains constant. The second is the idea of the independent bass voice which supports all the voices above it and exists in a state of continual tension and cadential relaxation with them, qualifying and even altering their effect.

Classical string quartet writing perfectly exemplifies the first of these developments, as, for example, at the beginning of the final allegro of Beethoven's first 'Rasoumoffsky' quartet, known as 'Thème russe' (Figure 8).

The tune is stated twice, first by the cello, and immediately afterwards by the first violin. The texture now complicates: the viola begins a third statement of the tune, but Beethoven throws the listener off by altering bits of it and adding imitations of it in the violin parts. The texture continues to complicate itself, with a corresponding rise of tension between the parts, which communicates itself to the listener, who may now perceive only swirling movement in place of clear-cut tune. This may result in a sense of being momentarily lost, with a consequent rise of anxiety. But all is well again when all the voices come together at the cadence, the tune reappearing. The extract may be interpreted as a sequence of tensions and relaxations, expressed in terms of precise and individually identifiable events as well as more complex accretions of instrumental voices into swirling masses, which per-

Figure 8 Beethoven, String Quartet in F major, Op. 59, no. 1, 4th movement

haps are not, or even meant to be, totally graspable in all and every detail. The exceptional listener will be able to give them back *in toto*; but this might prevent him feeling the appropriate rise and fall of tension.

Such transitions between simple and compound effect are not in the final analysis confusing, because the proficient composer leads the ear and prepares it for each successive situation by what we call musical logic. The listener is not excluded. Gestural signposts prepare him for what comes next, providing just enough for him to hold on to and not so much as to allow his mind to wander off into reveries. Regular phrase lengths and a regular rate of textural alteration secure him, giving him the necessary confidence to remain a participant, holding on to the background metre and, at one and the same time, perceiving foreground and relating it to background. The process has been variously described by psychologists and critics such as H. Keller and L. B. Meyer (in his book *Emotion and Meaning*) in terms of the frustrations and resolutions of aroused expectations.

An analogous (as it were) double perception is evoked where an inde-

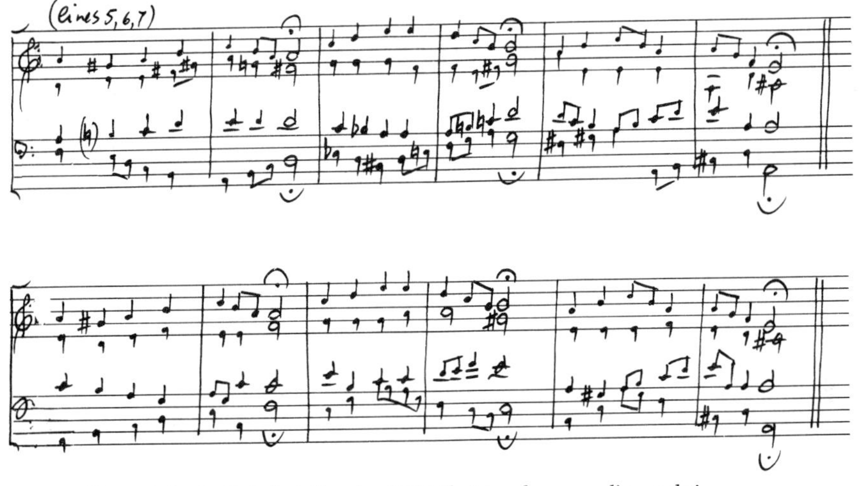

Figure 9 Bach, *St John's Passion* (12) 'Christus der uns selig macht' (2 versions, the second version transposed (no. 35))

pendently moving bass line is added to a melody. As an example, here is a simple chorale melody, once well-known (at least to Lutherans), easy to sing and remember, as set for voices and instrumental accompaniment in four parts, as it appears in the St John Passion (Figure 9). Bach's bass alters the effect of the chorale melody by stressing and colouring particular notes of it. The gaps or intervals between it and the chorale melody are filled in by middle voices, softening the tensions created between the two. The bass itself artistically alters the effect of the chorale melody by stressing and colouring particular notes.

There is nothing natural or automatic about the relationship of the bass and consequently of the harmony to the chorale melody. No one particular bass is implicit in the melody. Our example is only one solution, and it is significantly altered when the chorale is sung again later in the work. In the second version, the first four and last lines are identical with the first version, but the fifth, sixth, and seventh lines are altered. The need for this alteration probably arises out of the change of text (a different stanza of the same poem).

In a famous remark, Rousseau observed that for the common man a melody does not of itself suggest an independent bass, and surmised that if required to provide a descant to a melody, this common man would merely double it an octave lower, whichever suited his voice better. If this is in fact what goes on naturally, this common man will inevitably compare the composer's independent bass line with his own 'natural' doubling: once again implying a foreground–background relationship.

Musical meaning might just as well then not be something which is transmitted *by* organised sound but arise directly, or rather *be* the changing tensions and recolourings of background brought about by foreground elaboration.

'AVANT-GARDE MUSIC'

Not to understand, not to perceive meaningful continuity might then be no more than a failure to relate passing foreground events to background, with a resulting inability to identify tempo, harmonic tension and textural logic. In this sense, a great deal of music written in the last sixty years cannot be regarded as a straightforward continuation of Classical and Romantic music, neither in the way it is conceived, nor in the way it is to be listened to. Background—foreground perception is inapplicable here because, in reality, insufficient background is implied. Continuity is fragmented or constructed of events unrelated to each other, pitch succession is too complex to be memorable, and constructional procedures too difficult to be perceived as aural logic. Composers are said to re-invent their musical language anew for each composition they write.

In such circumstances a sense of expectation, or feeling for good continuation and satisfactory closure, previously considered the necessary conditions for musical reception, can hardly be aroused. Easiest would be to assert (and there is no shortage of those who have done so) that such music has simply gone off the rails, and there are plenty who will believe in a kind of Spenglerite version of history, where great civilisations are in a continual process of becoming, flourishing and declining. Measured against the certainties of the not-so-distant past, the present with all its inherent difficulties (they might conclude) testifies to a decline of potency. At this late stage, I shall not consider such apocalyptic views, preferring to restrict myself to a few simpler observations.

Avant-garde music ought not to be dismissed out of hand, merely because it cannot be understood obviously as a continuation or extension of past music. Certainly some vestiges of past music are retained in it: for example, most of it is (more or less conventionally) notated and performed in a modified but recognisably traditional manner. Consequently it looks as if it might be understandable in more or less the same way as the music of the past has been understood. Observation of the way audiences react and what is said about avant-garde music suggest that listeners perceive it as a proposed continuation of the manner of earlier music, but failing to obtain similar pleasure from it, feel they have been cheated. Underlying coherence is presupposed to exist and is then not perceived: unresolved anxiety ensues. The difficulty arises from the frustration of a conventional expectation. The gap between sender and recipient is at its widest here.

Faced with this, composers have two alternative courses of action available to them. They can try to fight for a foothold in the traditional musical world, matching up their imaginative flights with observable and presumably learnable realities of past musical practice. But in fact the vast majority of them see this as a pointless, nostalgic and debilitating course of action, preferring in its place an experimental, exploratory activity. There is no way forward, they believe, but to start again from scratch: which would seem to mean abandoning the criteria of conventional logic or appropriate continuation in favour of the presentation of momentary, unrestricted and unrestricting sound-events. Uncombined signs and gestures, kaleidoscopic alterations of timbre and texture, aperiodic rhythms and freely hanging (quasi-expressive) phrases, fragmentary in origin, are densely presented so as to engulf the willing listener. Where earlier composers intended to communicate a particular aesthetic impression, and to do so aimed at clarity, subjugation of detail to broadly moving melody and rhythm and a carefully graded relationship of certainties and ambiguities, avant-gardists prefer saturation and prolixity of musical phenomena, aiming so to kick over their traces and thereby create what might be described as a magical effect. This music is to be instantaneously perceived in a state of shock created by rapid alterations, or in dreamy states brought about by an apparently endless extension of constantly repeating and more or less identical patterns. This music is to physically surround the listener, and in this way remove the conventional gap between sender and receiver. Some believe it works best with the high loudness levels and high-quality speakers of recorded music, which extend (and sometimes transform) locations and distribution of sounds. The listener is then materially immersed in sound. Exclusive attention is not even demanded; the composer hopes to create a new community, even a world, rather than convey specific information.

Gruppen by Stockhausen, *Répons* by Boulez, and much of Cage's music are the most radical attempts once again to alter the conventional relationship of sender and receiver by physically enclosing them in a single continuous context. The majority of composers have neither the material possibility nor the will to go so far (it means rebuilding concert halls, expensive equipment or regrouping of players), but even where they restrict themselves to more conventional locations, and have their works performed alongside works of the past, they subscribe to the spirit if not to the letter of such ideals: adopting the new criteria and new values of such music, and trying to make them work within the old forms of transmission in the concert hall.

CONCLUSION

Underlying what has been said has been an assumption that regardless of cultural considerations, the rules of perception insofar as they are known to us are invariants. But the degree to which they apply to each and every listener might vary and the reasons why they vary affect the quality of the listener's perception of the music. Categories of differentiation continually alter, even within a single society, which like our own is subject to rapid and simultaneous changes and accretions. Effective communication might seem to presuppose an identity of attention, and a body of common expectations, the possession of which puts the communicants in context and ensures some authenticity of personal experience. Because, of course, we believe that there are authentic and inauthentic ways of understanding and different communities imply different authenticities.

As far as is possible, we now make a point of defending such authenticity as remains to us. Yet the intention to preserve it seems to run counter to that very openness to new experience which appears to be the only way of coming to terms with the very variety of currently available cultural manifestations. So extensive is this variety, that it might almost seem preferable to try to destroy what remains of any feeling for authenticity and context. To do so would be as it were to de-regulate musical experience – all musics: Mozart and Machaut, Stockhausen and Stravinsky, Gregorian chant and the music of the Gambias would be no more and no less than potentially evocative noise to be turned on and off as, when, and how required...

There seems something rather threatening in such an invitation to clear the mind and leave it blank, so that new characters might be drawn upon it.

8

Communication and technology

JOHN ALVEY

It is a daunting task to give the last lecture in a series as prestigious as this one. It is nearly two years since I was invited to lecture, and nearly three years since I left British Telecom, and so I am grateful that the title of my talk is ageless. Technology rolls on at an ever accelerating pace, and I am almost certainly out of date, but the principles of how to apply technology do not change much.

I am going to discuss five principal topics in this lecture: telephones and switching, optical fibres as a new form of communicative channel, the use of moving pictures in telephone lines, radio and mobile telephones, and the future technology of television. But I will begin with a few remarks about the applications of technology.

Any technology is irrelevant unless it is used by people, and in my experience the best return for effort comes when technologists live in the real world: the world of customers and marketeers, and even accountants, a necessary evil, who should be 'on tap but not on top'. And in this world the possibility of installing competing communication networks at a reasonable cost depends heavily on technology, a technology that is, as I've said, changing ever more rapidly.

Of course it is necessary to balance the options opened up by new technology – options that are sometimes far in the future – with a perception of how to exploit these options, granted that the market is often profoundly affected by government. But my experience again is that most direct marketing exercises depend on asking the customers what they would like: to which the answer usually is: 'More of the same, but better and cheaper'. That is not

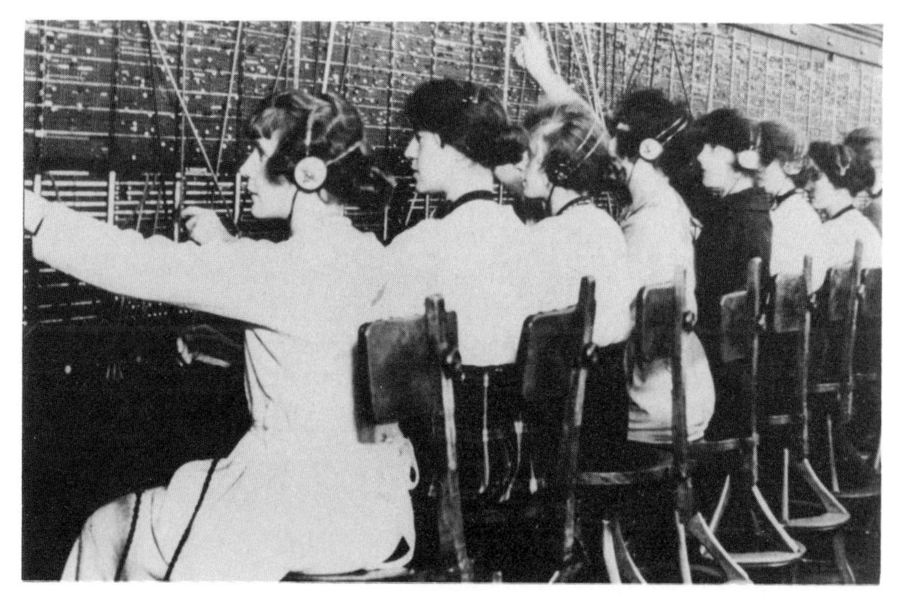

Figure 1 A manual exchange, 1922

at all a bad answer. Indeed, many successful companies have a marketing policy that consists in introducing a new technology which at first does a current task better than available services – and then they train the customer to accept and to use the new facilities embedded in that technology.

We all train our customers – of course we do – but we train them to be content for the moment with what it is economical to provide. And that is often far from ideal. The telephone is an excellent example. When Alexander Graham Bell invented the telephone, he had to use a microphone that responded only to the middle range of the frequencies that the voice generates. High fidelity sound ranges from a few cycles per second – perhaps ten – (or ten hertz, to use the correct terminology) to about 20,000 hertz (Hz). But Bell used a bandwidth of only about 4,000 Hz, because his microphone did not respond to low frequencies, and heavily attenuated higher frequencies. And telephone networks still use that same limited bandwidth, which transmits speech that is indeed intelligible, but distorted.

Now, of course, microphones have improved enormously during the last hundred years, but other parts of the telecommunications network are still limited to this narrow bandwidth: particularly the copper wires which carry the signals to your house. The technology now exists to give you much less distortion, but it costs money. You accept 4,000 Hz: how much more would you pay for a greater bandwidth?

Figure 2 A strowger switchboard

The pace of technological change in the field of communications is now so rapid that few of us can keep up with it. I can illustrate this point in terms of switching: how you route calls through a telephone network. Back in the 1920s, a telephone exchange was totally manual, and all calls were switched by the switchboard, as in Figure 1. If we had kept that system, you could have predicted confidently that as the telephone network grew, every lady in the land would have had to become an operator! The answer to that problem was, of course, automation. The automatic switching system is rather remarkable. It is reputed to have been invented by an American undertaker who was not getting any business because his rival's wife was running the local switchboard – so he invented an automatic system.

The system in Figure 2 is not quite what he invented, but something like it. It is slow, because mechanical switches are slow, and it therefore takes a long time to connect a call, particularly over a long distance where the call has to go through several exchanges. Modern electronic switches work far faster, but the replacement of a total network – 6,500 exchanges in the UK, for instance – takes time, and the need to make dialling speeds compatible with switching speeds means that it still takes a little time to connect some calls even through a modern telephone network.

The spread of automatic switching was not in fact all that rapid. The exchange illustrated in Figure 2 dates from 1953, when the UK network still used many manual switches, with only 4 million customers connected to it,

Figure 3 Part of Very Large Scale Integrated circuit (VLSI)

compared to more like 23 million today. The really important development in the 1950s was the coming of age of programmable digital computers. They were very much a thing of the present in those days, even though they were not powerful by present-day standards. But there was a strong convergence between the technology used for switching telephone conversations and the technology used in computers. It was effective and cheaper for telecommunications to go down this digital road, so that nowadays our current 'System-X' exchanges do not just look like computers: they *are* computers.

One technology that has driven this development is large-scale integrated microelectronics. The design of large-scale integrated chips has almost become an art form: indeed, when one magnifies such a chip it looks like a work of abstract art.

But this technology is already old: the chip in Figure 3 was made a few years ago. If you consider the dimensions of its fundamental components, then probably the smallest dimension is about five microns (i.e. five millionths of a metre), and the whole of that circuit could be covered by a human hair. That is rather small, and yet recent technology has improved on it. A more modern chip, such as the one in Figure 4, is a complete microprocessor,

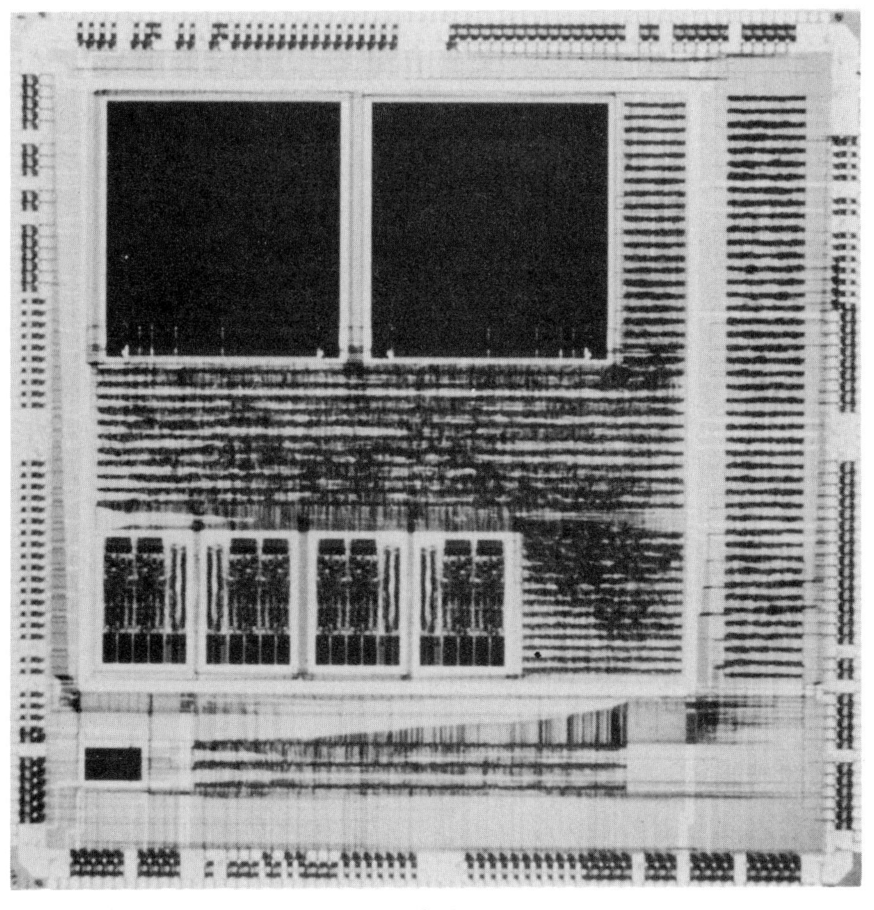

Figure 4 1750 computer on a single chip

with a quarter of a million transistors on it, and with the dimensions of its smallest components brought down to about one micron.

This development matters because the efficiency of modern computers is measured in terms of how many basic instructions per second they handle; and the 1750 chip in the picture handles about 4 million instructions per second, i.e. 4 MIPS. Put one more chip alongside it for handling memory and input and output, and you have a complete sophisticated computer on two chips. And one great advantage of such a very large-scale integrated chip is that the interconnections among the transistors are all on the chip and therefore the connections are very short. Hence the chip can perform very rapidly: it is not slowed down by the need for current to flow over relatively large distances, like those between the components of a TV set.

(One problem that cannot yet be avoided, however, is input and output. I have not counted how many pins there are on the chip in Figure 4, but it must be of the order of 150; and all of those pins have to be straight, because otherwise they will not plug into the circuit which you are trying to make.)

And all this power has now become quite cheap. People who use Sun work-stations may know that their central processor, which is a really powerful one, works at 8 MIPS, and yet it can now be made for 79 dollars: about ten dollars per MIP. That price is beginning to be very competitive, because it is a tenth of what the previous technology achieved. The reason, in part, is the much greater yield of satisfactory chips. They are made from large sheets of silicon with an impurity content measured in parts per billion, and variations in the flatness of the surface down to a micron or two. Any dislocations in the crystal structure must be minute, because a chip is a centimetre square, and its circuit will not work if there is a dislocation in the middle of it. And yet nowadays 80 per cent of chips that are made are satisfactory.

But how, you might wonder, is such an incredible work of art designed? The chip in Figure 4 contains a quarter of a million transistors, and they have to be interconnected accurately and efficiently. Even to work out how to test a chip like that is an enormous task. The 1750 computer chip needed 24 million design tests to ensure it functioned correctly. Other designs need more. It would take a long time to sit down and design a chip by hand; it would take a long time to test a chip manually. Naturally, both design and testing depend on the use of computers of a very powerful format. Even for the computer-aided design of a Sun work-station chip, you need immense computational power. And yet, one can turn round the design of a chip in a matter of weeks, from the basic concept to the piece of silicon in the hand of the user. On occasions some firms have made design modifications to a chip over a long weekend.

This technology is available, incredibly powerful, and has tremendously far-reaching implications for telecommunications. It underpins the new systems of switching, such as the System-X exchanges, which offer the user all sorts of features that were hitherto unavailable. For example, you can now punch in all of a customer's data, including information about the particular calls that he wants to accept or to reject at his own number, and whether he wants to divert his calls from his house to somewhere else where, say, he is playing bridge. But the power of such signal processing depends on an enormous financial investment. You do not make a plant for constructing chips for much under £100 million.

Let me now turn to a second technology that is one of the great developments of the present moment: the optical fibre. One of the things that I find interesting has been its long gestation time. It is twenty years since Standard Telecommunications (STC) first suggested that the transmission of signals by optical fibres was feasible; and to make it a reality has needed heavy investment over those twenty years by Corning Glass, British Telecom, STC, AT&T, and many other companies – it was difficult to persuade an accountant that it was worth doing. Indeed I am still not sure the City recognises the need for such forward-looking investment, or wants to have an industry that invests on a twenty year timescale.

Why are optical fibres so advantageous? The answer depends on three factors: a small size, a lack of signal degradation, and an enormous capacity for carrying signals.

An optical fibre is about as thin as a human hair, i.e. 120 microns in diameter: but the special glass core in the centre, which actually carries the signal, is only about seven microns in diameter. That is truly minute: if I hold an optical fibre in my hand, it is almost invisible. Its capacity for communicating signals depends, as we shall see, on the fact that it uses light. It uses a part of the electro-magnetic spectrum just below the frequency of visible light in the infra-red region. Signals can be carried on optical fibres for an incredible distance before they have deteriorated or attenuated to the point where they need to be amplified again. That is very important because the 'repeater spacing' (i.e. the spacing between the points at which the signals have to be regenerated) can go up enormously. Look at Figure 5: an old one, but I make no particular apology for that, because it shows that we got our forecast right.

Figure 5 illustrates the increasing spacing between repeaters with the evolution from copper cables to optical fibres. Thirty kilometres without any amplification was an important distance for British Telecom. At the time we started installing optical fibre, we had nice brick buildings with mains power in them spaced out at not more than thirty kilometre intervals: and it is more economical to put electronics in these buildings than underground. Hence, thirty kilometres was an important target, and we reached it fairly easily – in 1981. We have subsequently improved on it. Of course, the fewer amplifiers you have to put in, the cheaper the whole system is; and as the figure shows, from London to Paris is feasible, with just a laser in London, a sensitive receiver in Paris, and a 7 micron diameter glass fibre between them. We have not quite achieved that target yet, but we are almost there.

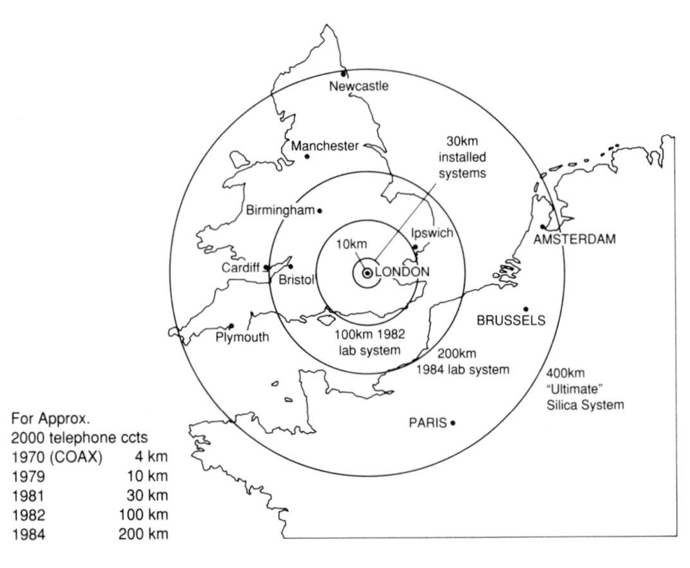

Figure 5

Perhaps the most important feature of an optical fibre is the tremendous number of signals that can be transmitted down it. We have seen earlier that a telephone conversation involves a 4,000 Hz range of frequencies, perhaps 1,000 Hz to 5,000 Hz. If one sends a second conversation along the same wire, there will be serious interference between the two if the same frequencies are used. However, if one adds, say, a constant 5,000 Hz to the second set of frequencies, then one can send two signals, one covering 1,000–5,000 Hz and the other covering 6,000–10,000. They will not interfere with one another, and can be divided into separate signals at the receiving end by suitable filters, after which deducting 5,000 Hz from the second signal will reproduce the original 1,000–5,000 Hz conversation.

For this to work, of course, the wire must allow the transmission of frequencies in the range 1,000–10,000 Hz rather than just 1,000–5,000 Hz. We put this by saying that its bandwidth must be 9,000 Hz rather than 4,000 Hz. And the greater the wire's bandwidth, the more this process can be repeated, and the more conversations it can carry at once. Optical fibres, operating at the very high frequencies of infra-red, can support enormous bandwidths, and therefore they can carry a vast number of signals. Look at Figure 6, bearing in mind its logarithmic scale: what it shows is that an optical fibre is operating at about 10,000 times the highest frequency used in current radio transmissions. And since you can usually exploit about a tenth of the available frequencies at the centre of the band – things get non-linear outside

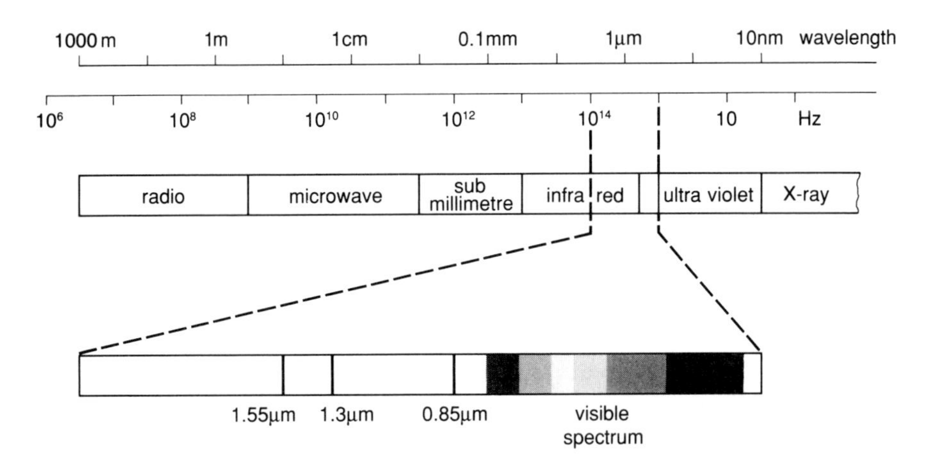

Figure 6

that range – one single optical fibre can transmit about a thousand times more signals than can be transmitted over the entire radio bandwidth.

Moreover, because radio waves are broadcast through the atmosphere, a given frequency can be used only for a single signal without risk of interference. But if signals are constrained within an optical fibre, then the same frequency can be used for different signals without risk of interference merely by transmitting the signals down separate fibres. So if a thousandfold increase in the capacity of radio waves is not enough, you merely have to put another optical fibre in your cable.

There are something like 600 million telephones in the world. None of them is in use all the time – though if you have children you may not believe that – and so there are perhaps 350 million telephone conversations going on at any one moment. Every one of them could be carried by a single optical fibre. It would not make a very good network, but it could transmit them all.

The amount of information that can be transmitted depends directly, as we have seen, on the bandwidth of the communicative medium. Television, of course, conveys very much more information than the telephone, and so you cannot transmit through a particular medium as many television pictures as telephone calls. Even so, you can transmit digitally 50,000 high-definition television signals down one optical fibre, and you do not have to waste any of the radio spectrum, which is after all a very precious part of the electromagnetic spectrum: they are not making any more of it.

The virtually unlimited bandwidth of optical fibre has another important consequence. Once a fibre is installed, you are future-proof. No subsequent technological development in communications will require you to rewire the links between users. There is already a lot of optical fibre in the ground, including thousands of kilometres of it in the UK. Some has been installed by British Telecom, some by Mercury, and some by other companies. You can install it rather like you run a ring main round a house and tap off your electricity wherever you want it. It is similarly quite possible to run a 'ring main' of optical fibres with its enormous bandwidth round a whole district, and to tap off outlets from it wherever signals are needed.

There do, however, remain one or two technical problems to be solved before the system can be perfected. One is the manufacture of lasers of appropriate stability. The bandwidth for a signal depends on what the signal is to transmit. The maximum bandwidth now is about 2,000 mega-hertz (MHz), or equivalently 2 gigabites per second (a giga equals a billion, i.s. 10^9), which allows a signal to transmit a vast amount of information: every half a second, the equivalent of the entire contents of the Encyclopedia Britannica will land in your in-tray. (Whether you want all this information is an interesting question, to which I shall return.) And you can get something like 13,000 two gigabit channels down one optical fibre – provided that the frequencies can be separated accurately enough. And that, in turn, depends on the manufacture of sufficiently stable lasers. The stability that has so far been achieved is of the order of about one part in a billion: enough to provide 1,000 channels in a single fibre, each of which, as I say, can carry an enormous amount of data.

There is a lot of argument at present about whether competitors should be able to use British Telecom's ducts for installing their own competing systems. Ducts are the expensive part of putting cables in the ground: you have to dig up the roads – nobody loves you for that – and it is in fact a costly operation. When the road is dug up, you install pipes, like water pipes, and then drag cables through them. To allow future network expansion, the pipes usually have spare capacity: and that is what new service providers would like to use in order to reach their customers at an economic cost. There are practical difficulties in doing this, of course, but perhaps the arguments become irrelevant if an optical fibre ring main is installed everywhere. For then, with such a system, any new service provider could easily find spare bandwidth with which to reach his customers.

And how do you install optical fibre? It is as thin as a hair, and if you try to pull it like you pull an electrical cable, naturally it snaps. So our cables

Figure 7 Submarine cable-laying ship

are assembled with a strength member up the middle, and you pull on that, which pulls the fibres with it. But if you want to install a single optical fibre (and you might well) then there is another way, which was discovered by accident at the British Telecom labs at Martlesham in the days when I ran those labs: the fibre can be *blown* through a fairly narrow pipe. At first we thought we would try putting a little parachute on the end of the fibre and blowing the parachute through. It worked like a charm: the parachute went through the pipe, dragging the fibre nicely behind it. But even after the parachute popped out of the end of the pipe, the fibre itself kept on going and going. Serendipity is a great thing: the friction of the air on the glass fibre itself moves it though the pipe, even around quite sharp corners.

Optical fibres now cross oceans. Cable-laying ships lay them down along with other cables under the sea. The first transatlantic optical fibre was opened last year; and optical fibres will eventually circle all the major oceans, including the Pacific, now that one unanticipated problem has happily been solved: for some reason sharks like them and tend to bite through the cables!

Transatlantic signals have to be regenerated, of course, because it is much more than 400 kilometres across the Atlantic. So there have to be regenerators on the ocean bed, and it is extraordinarily expensive to go out into the Atlantic to replace one if it fails. One of the interesting features of typical regenerators therefore is that the people who make them have to guarantee them to work for twenty-five years. That is quite a tough guarantee: not many cars, for instance, are guaranteed to run that long. And if one of them does fail, then the manufacturers have to go out to mend it – and pay for the loss of traffic while it is being repaired.

What are the implications of what I have been saying? For instance, I have talked about the fact that optical fibre has an enormous bandwidth, which yields a massive communication capability. But do people want that kind of capability? Can they handle data that comes in at that sort of rate? It has been suggested that perhaps so-called 'expert systems' can help to sort the wheat from the chaff in such vast amounts of data. (Expert systems are computer programs that contain large sets of rules embodying human expertise, and they already exist to help human users to reach informed decisions about where to drill for minerals, for example, or how to configure a new computer.) But I have my doubts. I remember Roger Needham, the Professor of Computer Science here in Cambridge, talking on television to a visiting professor from the Cambridge on the other side of the Atlantic, and asserting that for a very long time expert systems will have the IQs of four-year-olds. The man from Massachusetts was more optimistic; but Roger is probably right.

Yet there are already one or two uses for the massive bandwidth that optical fibres can provide – at least for computer-controlled processes. For example, Volkswagen have manufacturing plants in Germany that are controlled from design houses in Berlin, and the total design of a car has to be transmitted extraordinarily fast. A design contains an immense amount of data, and the company is using an experimental optical fibre system to transmit it.

But perhaps the prime way in which this kind of bandwidth will be used is for the transmission of moving pictures. Television pictures have something of the order of 500,000 picture points on the screen, which have to be scanned once every twenty-fifth of a second, specifying the light intensity, and colour, of each point. And that is enough data to use up a significant proportion of the bandwidth.

The communication of moving pictures makes possible meetings between participants who are geographically remote from one another. Each person has a confra-vision studio in which such meetings can occur. You can look at and talk to any of the people taking part, even though they might easily be on the other side of the Atlantic. And my experience of this form of social interaction is that you rapidly forget about the medium that sits between you and the people at the other end of the transmissions. The meeting is entirely normal, and the medium works very well.

Figure 8 Confra-vision studio

Of course, even now we have all this bandwidth, people still work hard to use less, because it costs more to use more. And a pertinent fact about moving pictures of meetings between people is that what move most – provided that they are not playing musical chairs – are their mouths as they speak and their hands as they gesticulate. So it is necessary to keep transmitting only those bits of information that encode the parts of the picture that are changing. The prime picture need be transmitted only initially, and can thenceforth be regenerated locally. Hence, the pictures needed for a remote meeting can be transmitted with a considerable reduction in data flow.

Whether this procedure is worthwhile, of course, is a matter of cost. Normally one transmits television pictures with a bandwidth of five MHz. Hence, if the signal is transmitted digitally, it is equivalent to 100 megabits per second, whereas much confra-vision is done at a rate of only two megabits per second. Indeed, these techniques of data compression have been pushed far enough to get the rate down to 64 kilobits per second, naturally with a corresponding loss in picture quality. At even lower bandwidths it would be possible that the only information transmitted is 'Granny is now smiling', and a still picture of Granny at the other end slowly changes

Figure 9 Financial dealer board

to a smile. Whether the customer will think the images produced in this way are worth it remains to be seen.

Another place that uses a great amount of data is a dealer desk in the City, where they handle currency exchanges, and really do have to cope with continually changing data. And the decisions they make, of course, involve millions, or tens of millions, of pounds. They can lose a lot of money rather fast if they make the wrong decision. It is a tough environment, and dealers tend to burn out at forty. At these desks, they can access data at a vast rate from various sources around the world. In the one shown, there is a 'touch screen' – the dealer just touches the point on the screen that gets him the service he wants. A typical dealer wants immediate and continuous access to about twenty data bases, which are part of the sixty or so that he might need to access. Each of these data bases is very expensive to produce and maintain, and the companies who produce them charge a lot for access to them. Dealers do not need a continuous access to them all the time; and the fee is usually charged on the basis of a payment for each time a consultation occurs. The ideal system is therefore one that updates the information from these data bases rapidly but rarely. Rapid updates demand a large bandwidth; but rare updates use the bandwidth inefficiently. Yet, if not using the bandwidth means the loss of a lot of money, then customers will pay to use it.

But what we all expect to use the great bandwidth of optical fibre is television: the Indians will come down through the Grand Canyon. Television makes such an obvious demand on bandwidth that many people have felt that it will drive companies to invest money in broadband systems. I suspect that the business systems will drive it first, and then the television systems, which will certainly not be far behind. And one already sees satellites used on this basis. Not long ago, I was wandering around Wyoming, where there are probably fewer people per acre than anywhere else in the Western world, and I came across a lonely house with a TV satellite dish. There was a man sitting outside his family 'one holer' watching television. That, perhaps, epitomises communication in the modern world.

At this point, I want to return to radio. There is always a problem of balance, as I've said, in the use of the radio spectrum because it supports just one service per area within a given bandwidth. Does one therefore use the spectrum for mobile radio transmitters that have no other option, because there is no other way for mobile units to communicate? Or does one use it, as it is intensively used in this country, for television coverage? In this country there are very large installed bases of television transmitters, which have in fact just been renewed. They provide an excellent television coverage, but they occupy frequencies that one might well want to assign to mobile transmitters. Cellular transmitters, for example, of the sort needed by car phones are very cramped at this moment, and there are nowhere near enough channels for them within the bandwidth allocated by the government. Hence, whenever there is a hold up on a motorway, such as the M25, then everybody with a car phone who is stuck in the jam wants to use it. The result, of course, is that nobody makes a call: the system collapses.

One crucial point about the cellular phone is that it is used from a moving vehicle. The system must therefore allow for the movement of the vehicle: it cannot just allocate a frequency to that vehicle and allow it to use that frequency anywhere. Instead, the country is divided into cells, and each cell has only a limited number of channels: otherwise the interference problem would become enormous within the limited bandwidth that is allocated to cellular phones. So the cells are not very big – just a few kilometres across. And as a vehicle moves from one cell to the next, somehow the system has to recognise the move, cancel the vehicle's call in the cell it is leaving and reinitiate the call on a different frequency in the cell it is entering. That is an easy enough process to describe, but it is a difficult job to do, and it takes a lot of expensive apparatus. At the same time, the system also has to do a few other things – like making sure that the bill goes to the right person. I mentioned the M25 because it is a place where severe difficulties occur for this kind of apparatus. There have been heroic attempts to solve these problems, to reduce the cell sizes and so on. They cost money too.

Cordless telephones are useful also, particularly for those of us who are not as mobile as we might be. These are simple phones that make radio contact with their own base stations. There are a limited number of frequencies allocated to them. Their transmitters use only a very low power, and so the transmissions do not go very far. Hence the same frequency can be used nearby. A new form of cordless phone, which British Telecom and others will soon be introducing, is called 'phone-point'. The phone fits into your inside pocket, and it has its own batteries and transmitter. It is low-power, and so the batteries are small. There will be a base station like the base station for a cordless phone to which you make radio contact. But about forty different people can use one base station, all on different frequencies. When you want to make a call you go to a place that has a phone-point: it will have a 'Phone-point' sign and probably some advertising. You can make a call using your vest-pocket device through the phone-point into the basic network.

What phone-point cannot do, at the moment, is allow you to receive calls. The system will not call you, because you may be out of the range of the base station. But there are ways around this difficulty. One is the pager. With this device, if I want to contact you, I phone the number of your pager. My call goes into a queue, and there may be a four-minute wait until it reaches the head of the queue. When it does, the particular pulse code corresponding to your pager will be transmitted. Your pager then bleeps, and when it does you respond in a prearranged way – perhaps you phone your office. More sophisticated pagers can be made: one can send messages to them, and the code that reaches them is long enough to have, say, a phone number on it. You can then use your phone-point to make a call to that number. Of course, there may be a four-minute gap between the person first trying to contact you and you receiving the message and going to the phone-point. Nevertheless, this device, I suspect, is going to be very powerful and very popular. There will shortly be between 500 and 750 stations that will accept this device within the inner circle line in London. The calls should be about the same price as from a phone box. You have to buy the vest pocket phone, but it will not be expensive. Very large-scale integration, i.e. the microchip, will enable it to be very cheap.

Finally, does this country need a broadband network using optical fibre for much of its transmission? And if so, what should the government's role be in setting it up? The committee, on which I served, that studied this problem looked at three options. The first was that the government should push for a broadband network as soon as possible. The second was that the government should do very little, apart from remove most of its currently inhibitory regulations, so that entrepreneurial competition could reign supreme. The third option was somewhere in the middle: there should be no government subsidy, but there should also not be a complete free-for-all.

Perhaps because politics is the art of the possible, we came down somewhere in the middle, for the third option. It seemed most unlikely that a subsidised push for the first option was commercially justified. Likewise, it seemed that the second option, the free-for-all, was unlikely to produce anything of lasting utility. We felt that a broadband network would arrive fairly soon, so long as people investing in it could have reasonable hopes of returns. The actual types of services, however, and the volume of their use, are not really clear to anyone, and will probably take us all by surprise.

But how does one do the necessary research and development, which will certainly be both difficult and expensive? A European programme, which is aimed at sharing the costs and risks amongst the various European industries, has made some progress in generating working relationships and relevant technology; and a strategic audit concluded recently that the programme should now concentrate more on what customers are likely to need – largely so that network operators will have a better idea of how to sell broadband facilities when they install a network. But one way or another, I am confident that the broadband network will arrive: it is, as I said earlier, the ideal medium to get any information to any user, whether fixed or mobile.

One particularly fascinating service that it will enable is high-definition television. Some people will remember black-and-white television. Those sets had about 250,000 picture points on the screen; but with the move to colour the number of picture points doubled to about 500,000, and the resulting improvement in picture quality was highly marketable. High definition television puts perhaps four times as many points on the screen, and the eye can no longer resolve individual spots on the screen as it can now.

You can therefore be close enough to the screen – or the screen can be large enough – for the sides of the picture to appear in our peripheral vision. Indeed, if the picture were stretched a bit sideways, altering the aspect ratio from 4 to 3 to 5 to 3 (which is close to the Golden Ratio), the effect is even further enhanced. One begins to be immersed in the picture. There may even be a distinct three-dimensional effect, which at present one can get only in a few very expensive presentations on three-dimensional screens, about the size of a house, and using a large number of projectors. But the result is indeed an intense three-dimensional effect – you are actually sitting on that raft going down the Grand Canyon.

But technology of this sort has to have standards that are always upwards compatible. A new system of television must generate pictures that can be received on current receivers, but which can also give an improved performance on a new generation of receivers: and so on up until one reaches the Holy Grail of high-definition television. That way, of course, one always has a substantial audience.

Let me finish with two visions. One is of a country connected by two-way high-definition television, with as many channels available as necessary. You can go to any football match while sitting in your arm chair, and your holographic image can be at the ground, cheering the goals or shouting at the referee, but not invading the pitch. If the game disappoints you, you can leave at the flick of a switch.

The other vision is of a totally mobile telephone system, where my pocket phone is me, and my telephone number, when it is dialled, always finds me wherever I am in the world. My telephone number will be derived from the genes of my parents, and will be written on my forehead. So that at least the numbering system will be controlled by God – and not by the Department of Trade and Industry.

ACKNOWLEDGEMENTS

Jacket David Redfern

CHAPTER 1

Figures 1 and 2 were prepared by Horace Barlow.

Figure 3 and Figure 7 From *Eye, Brain and Vision* by David H. Hubel. Copyright © 1988 by Scientific American Library. Reprinted by permission of W. H. Freeman and Company.

Figure 4 From H. B. Barlow, R. M. Hill and W. R. Levick, 'Retinal ganglion cells responding selectively to direction and spread of motion in the rabbit', *Journal of Physiology*, 173, 377–407 (1964).

Figure 5 From C. W. Oyster and H. B. Barlow, 'Direction-selective units in rabbit retina: distribution of preferred directions', *Science*, 155, 841–2 (17 February 1967).

Figure 6 From J. J. Simpson, 'The accessory optic system.' Reproduced, with permission, from the *Annual Review of Neuroscience*, vol. 7. © 1984 by Annual Reviews Inc.

Figure 8 From W. R. Levick, 'Receptive fields and trigger features of ganglion cells in the visual streak of the rabbit's retina', *Journal of Physiology*, 188, 285–307 (1967).

Figure 9 From C. Blakemore, 'The baffled brain' in R. L. Gregory and E. H. Gombrich *Illusion in Nature and Art* (1973). Reproduced by kind permission of Gerald Duckworth & Co. Ltd.

Figure 10 By courtesy of Professor Janos Szentágothai.

Figure 11 Adapted from Hubel and Wiesel 1977, and Hubel 1988.

CHAPTER 2

Figure 1 Ian Wyllie.

Figure 2 From drawings by P. Leyhausen (cat) and P. Barrett (wolf).

Figure 3 Kim Taylor/Bruce Coleman Ltd.

Figure 4 By courtesy of T. Clutton-Brock.

Figure 5 From *Owen: Wildfowl of Europe*, by kind permission of Macmillan, London and Basingstoke.

Figure 6 Bill Wood/Bruce Coleman Ltd.

Figure 7 L. Barden

Figure 8 David Linden

CHAPTER 4
Winnie-the-Pooh:

From *Winnie-the-Pooh* by A. A. Milne, illustrated by Ernest H. Shepard. Copyright 1926 by E. P. Dutton, renewed 1954 by A. A. Milne. Reproduced by permission of the publisher, Dutton Children's Books, a division of Penguin Books USA Inc.

From *The House at Pooh Corner* by A. A. Milne, illustrated by Ernest H. Shepard. Copyright 1928 by E. P. Dutton, renewed 1956 by A. A. Milne. Reproduced by permission of the publisher, Dutton Children's Books, a division of Penguin Books USA Inc.

Line illustrations by Ernest H. Shepard, copyright under the Berne Convention, reproduced by permission of Curtis Brown, London.

From *Winnie-the-Pooh* and *The House at Pooh Corner*. Used by permission of the Canadian Publishers, McClelland and Stewart, Toronto.

CHAPTER 5
La Liseuse de roman by Vincent van Gogh by courtesy of Marlborough Fine Art.

CHAPTER 7
Photo of Oscar Peterson and company: by courtesy of Dennis Stock, Magnum Photos Inc.

CHAPTER 8
Pictures by courtesy of British Telecom.

INDEX